A Cabinet of Medical Curiosities

A Cabinet of
Medical Curiosities

❧ Jan Bondeson

W. W. NORTON & COMPANY
New York • London

Library of Congress Cataloging-in-Publication Data
Bondeson, Jan.
    A cabinet of medical curiosities / Jan Bondeson.
        p.   cm.
    Includes bibliographical references and index.

    1. Medicine — Miscellanea.   2. Medicine — Anecdotes.   I. Title.
    R706.B66   1997
    610 — dc21   97-20749

ISBN 0-393-31892-3 pbk.

W. W. Norton & Company, Inc.
500 Fifth Avenue, New York, N.Y. 10110
www.wwnorton.com

W. W. Norton & Company Ltd.
10 Coptic Street, London WC1A 1PU

        5 6 7 8 9 0

Contents

Preface

In the seventeenth and eighteenth centuries, natural history repositories and medical museums still resembled the old-fashioned cabinet of curiosities. In these kaleidoscopic displays, the most curious and dissimilar objects were exhibited side by side: a dried mermaid, the shin bone of a giant, a unicorn's horn, Egyptian mummies, and African natives' gear were intermixed with pickled and dried monstrosities, the skulls of hydrocephalics, and the death masks of famous men. The layout of this book resembles that of one of these old medical cabinets of curiosities, rather than a conventional textbook on medical history. It aims to depict the odd, the bizarre, and the unexpected.

In 1986 I first had occasion to read Gould and Pyle's *Anomalies and Curiosities of Medicine*. This weighty tome of 968 pages, originally published in 1897, contains a fascinating overview of the darker side of medicine. It seemed considerably more interesting than the dry-as-dust textbooks I was reading at the time, as a final-year medical student. With gusto as well as learning, Gould and Pyle discussed strange diseases, remarkable malformations, uncommon and gruesome ways of death, and unlikely feats of fasting or gluttony. In 1726 a country lass from Godalming in Surrey, Mary Toft, convinced King George I, the Prince of Wales, the court anatomist Mr. St. André, and a host of the learned and the ignorant that she had given birth to seventeen rabbits. Her attendants explained this extraordinary phenomenon as the result of a "maternal impression." Her longing while pregnant for a meal of delicious jugged rabbit was supposed to have brought about sinister changes in her organs of reproduction. Other phenomena discussed by Gould and Pyle were the colossal Irish Giant, the hairy "Apewoman" Julia Pastrana, and the marvelous Two-Headed Boy of Bengal, who had another living head, upside down, on top of his own. Some of the subjects discussed had more than a hint of the macabre: colonies of snakes and frogs living in the human stomach, kings and emperors who had been devoured by lice while still living, and the old horrors of premature burial and spontaneous human combustion.

Drs. George M. Gould and Walter Pyle had been involved in the research for the Index Catalogues, the great late nineteenth-century bibliography of the contents of periodicals in the U.S. Surgeon General's Office. Although no longer up-to-date, these Index Catalogues are still a valuable research tool for medical historians, and they provided a natural starting point when I started my own research into some of the subjects discussed by Gould and Pyle. In many instances, it seemed desirable not to limit the discussion to merely a medical point of view but to bring in aspects of ethnology, folk medicine, and literary and cultural history as well. The modern computerized Citation Indices of medicine, science, and arts and humanities, among other databases, have proved invaluable in finding new references and updating the research.

At the time of Gould and Pyle, the possibility of people combusting spontaneously was still debated, and the majority of both doctors and laymen in the United States still believed in maternal impressions. Even today, maternal impression lore is by no means forgotten, and the reality of spontaneous human combustion has recently been asserted in several popular books and TV programs. Even the ludicrous "bosom serpent," which has to be enticed out in some way after having taken up permanent lodgings as a parasite in the human stomach, has been resurrected among the modern newspaper canards. These phenomena provide some of the most striking examples that old medical errors die hard.

Sometimes, the opposite has proved true: modern research indicates that the legendary "lousy disease," which has even been used to exemplify the unreliability of the old chronicles of medical marvels, may actually have existed, although it would have been caused by mites and not lice. Advances in teratology enabled me to determine that the Two-Headed Boy of Bengal was not a unique freak of nature but could be classified with the parasitic conjoined twins. In the case of Julia Pastrana, whose pathetic career as a sideshow attraction continued for more than a century after her death, I have actually been able to examine her mummy and to have hair samples from her beard and whiskers and radiographs of her dentition analyzed in search of an accurate diagnosis.

I thank Professor A. E. W. Miles, London, for valuable cooperation in investigating the diagnosis of Julia Pastrana; he also made useful comments on several of the other essays. Mr. Arie Molenkamp, Loosduinen, has my thanks for finding several useful references on premature burial and translating extracts of them from the Dutch. I am also grateful to

Professor Birger Bergh, Lund University, Sweden, for checking Latin translations throughout.

The Crafoord Foundation, Lund, Sweden, provided several years of generous financial support for my historical research; without it, this book would not have been written, and I am grateful to the foundation. I also thank the Swedish Society of Writers for a generous scholarship that enabled me to work full-time on this book for several months in late 1995.

JAN BONDESON

London

A Cabinet of Medical Curiosities

Spontaneous
Human Combustion

 *So she was burnt, with all her clothes,*
And arms, and hands, and eyes and nose;
Till she had nothing more to lose
Except her little scarlet shoes;
And nothing else but these were found
Among her ashes on the ground.

Dr. Heinrich Hoffmann,
The English Struwwelpeter, 12th ed.

I N 1635 THE BOARD OF THE UNIVERSITY OF
Copenhagen was called in to give judgment on a difficult forensic
question. A Danish peasant named Peder Pedersøn had expired, and
the cause of death had been given as ingestion of excessive amounts of alcohol. This finding was apparently challenged by some learned judges, who
did not believe that it was possible for any person to drink himself to death
without self-combusting. The verdict of the university professors, signed and
sealed on December 19, 1635, was that it might well have been possible for
a man to die of alcoholic excess even though the flame of spontaneous combustion had not been seen to issue from his mouth and his lips were not
burned by the fire.

Nineteen years later, Professor Thomas Bartholin, the famous Danish

1

anatomist, discussed the problem of spontaneous human combustion. Bartholin, who discovered the circulation of lymph, was one of the leading anatomists of the seventeenth century and an early supporter of William Harvey. He also wrote a monograph on the unicorn and its medicinal uses, however, and in the somber Latin pages of his collection of six hundred curious anatomical findings and case reports, the most astounding tales may be encountered. Bartholin wrote about childbirth by the mouth, Siamese triplets with animal heads, and a cow pregnant with forty puppies. To put it mildly, he had a credulous bent, and his great interest in monsters and strange events sometimes got the better of him. Thomas Bartholin was also a firm supporter of the theory of spontaneous human combustion. It was contrary neither to experience nor to reason, he maintained, that excessive quantities of strong liquor could ignite in the stomach, and he gave several case reports. One of these had been sent to him by his old friend Professor Moreau in Paris. In another, from an obscure work by Everhard Vorstius, a Polish gentleman, living during the reign of Queen Bona Sforza, had drunk two large bottles of "brandy-wine"; suddenly, he vomited flames and was burned to death. The elder Sturmius described the case of three Courlandish noblemen at a tremendous "stag party," who drank until they were almost unconscious. All at once, flames burst from the stomachs of two of them and they were killed. The fate of the third nobleman is unknown, but if he survived both the risk of combusting spontaneously and the subsequent hangover, he is likely to have become much more temperate in his habits.

Three years later, Thomas Bartholin described yet another spontaneous human combustion, this time from his native Copenhagen. A scholar had been sitting at a tavern all night drinking strong vodka with his friends, not ceasing until a flame issued from his mouth with such force as to make speech impossible. His drunken friends managed to extinguish the flame by pouring water on him, and the man survived. He was rather frightened by this gruesome incident and decided to walk home to his professorial mansions on a nearby street; on the way there, he suddenly dropped dead. In his comments on the strange death of this sottish professor, Thomas Bartholin added that he had often seen gangs of drunken peasants settled outside the walls of Copenhagen, always lying on their backs with their mouths open to allow the spontaneous flame to erupt from their mouths unimpeded, so that it would not suffocate or burn them.

In 1673 Bartholin's nephew Oliger Jacobaeus described another case of the fiery death. An old woman in Paris, so addicted to strong liquor that she had consumed little else for two years, was found thoroughly incinerated. Only her fingertips and the cranium remained.

❧ The theory of spontaneous human combustion was incorporated into contemporary medical knowledge without any apparent objections. That a drunkard could be consumed by flames ignited within his own body did not appear unlikely to early eighteenth-century medical men, who were willing to accept the most bizarre hypotheses. The courts, too, were sometimes convinced. In February 1725 a most intricate legal process took place in the ancient French city of Rheims. The corpse of the wife of Jean Millet, the landlord of the Lion d'Or tavern, had been found smoldering in the kitchen; only the legs, part of the head, some dorsal vertebrae, and other large bones could be identified; the rest of the body was a blackened, fatty mass. The Rheims police arrested M. Millet and charged him with murdering his wife and setting the corpse on fire. It was well known that Mme Millet was past her prime and that M. Millet had cast his eyes on a comely young serving girl from Lorraine. She had lately been employed about the tavern, since Mme Millet was generally too drunk to do much work, even during the day. Although Jean Millet was one of the most respectable townsmen of Rheims, he was sentenced to death. Then one of his lodgers, the young surgeon Louis Nicholas Le Cat, saved him. Le Cat, well read in the recent medical literature, pointed out to the judges that Mme Millet had, in all probability, combusted spontaneously. Jean Millet was released, but the grueling process and the threat of execution had broken him. The wretched publican spent his remaining days in a lunatic asylum.

On the morning of March 14, 1731, a chambermaid went to wake her mistress, Countess Cornelia Bandi, in her town house in Cesena, Italy. The countess did not respond when her name was called, and her state bed was empty. The horrified servant saw, on the floor, four feet from the bed, a heap of ashes containing the countess's legs with the stockings intact, as well as her skull and three blackened fingers. None of the bedroom furniture had burned, not even the bed and bedsheets, but some candles on the bedside table had melted. The room was filled with a greasy, foul-smelling soot, and the floor was thickly smeared with a moist, sticky substance. The bottoms of the windows were covered with a yellow fluid of the most loathsome smell and appearance. Beside the heap of ashes was a small oil lamp, with no oil in it. A piece of bread in the pantry was black with soot, and several dogs refused to eat it. The countess, a lady of sober and temperate disposition, regularly used to bathe her entire body in camphorated spirits; this dangerous habit was of course blamed for her increased flammability.

The clergyman Giuseppe Bianchini, who described Countess Bandi's demise, added a curious note about the uncommonly flammable perspiration of the wife of one Dr. Freilas, physician to the archbishop of Toledo. Once, the erudite Padre Bianchini testified, her underwear burst into flames on exposure to the air "and shot forth like grains of gunpowder," though he does not say if he himself observed this remarkable phenomenon.

In 1744 Bianchini's account of the Bandi case was discussed before the Royal Society of London, together with another astounding instance of spontaneous combustion, this time from the town of Ipswich. Mrs. Grace Pett, an elderly fisherman's wife, had the habit of going down into the kitchen late at night, "after she was half undress'd, to smoak a pipe." She did so on April 9, 1744, and the morning after, her daughter found the smoldering remains of Mrs. Pett's body near the kitchen grate. When the glowing fire was quenched with water, the corpse presented a most strange and macabre appearance. The trunk looked like a heap of charcoal covered with white ashes, and the head, arms, and legs were also much burned. Mrs. Pett was not a habitual drunkard, but according to one account, she had consumed a plentiful amount of gin that night to celebrate her daughter's recent return from Gibraltar. The contemporary experts were at a loss to explain how she was burned to death, since there was no fire in the grate and the candle had guttered out in the socket of the candlestick beside her; a child's clothes on one side of the charred body and a paper screen on the other were both untouched by the fire. It was concluded that another spontaneous combustion had occurred: "The Manner of Fire burning in her Body is described as the workings of some inward Cause." Many of the townspeople believed that Grace Pett was a witch and that she had used her magic arts to steal sheep from a neighboring farmer. To rid him of her persecution, a local witch-hunter had advised him to burn the body of a sheep, and the same evening Grace Pett was consumed by the flames.

❦ In the early 1800s the belief in spontaneous human combustion was accepted throughout Europe. A great many very similar cases were reported, particularly from France and the German states: the victim's badly burned remains were found in a closed room, often but not always near the fireplace. The combusted individuals were often obese and elderly and usually female. At times, only some charred bones and a heap of ashes were left. An oily soot covered the furniture in the room, and a disgusting smell was sometimes noted while the combustion took place, usually during the night. Oddly, the corpse's feet, with the stockings and shoes undamaged, were occasionally found intact. Some believed that the victim's clothes could be unaffected, while the body was almost totally reduced to ashes within them, but this never really took place. Another aspect of the mythology of spontaneous combustion was that water did not extinguish the flames, but indeed fuelled them.

After Lavoisier and Priestley clarified the basic chemistry of combustion, considerable research was devoted to elucidating also the mechanisms underlying the spontaneous variety of this process in humans. Several doctoral theses were dedicated to this subject; the first of them written by Dr. Jonas Dupont, of the University of Leiden, in 1763. In the seventeenth and eighteenth centuries, most writers favored the simple hypothesis that alcohol saturation made the drinker's body flammable, but the French physiologist Julia-Fontenelle cast doubt on this theory when he tried to light lumps of flesh that had been immersed in alcohol and found that they would not burn. The German physician F. J. A. Strubel, who wrote a doctoral thesis on spontaneous combustion in 1848, suggested that electric currents arose within the body, reducing the body water to hydrogen and oxygen gas; the resulting, highly flammable oxy-hydrogen gas could easily be ignited by an electric spark. Another thesis writer, Dr. B. Frank of Göttingen, believed that the habitual toper's body became impregnated with phosphorus-containing compounds that rendered it flammable; this theory was received with much respect, and it was even accepted by Magnus Huss in his famous *Alcoholismus chronicus*, one of the classics of the literature on alcohol addiction. Some French physiologists claimed that a high fat content could make the body easy to ignite, others that "inner heat" could be engendered through static electricity when the components of the blood rubbed against one another.

The famous French surgeon Guillaume Dupuytren's hypothesis is a model of clarity beside these wild speculations. He denied that alcohol enhanced combustibility. Rather, it rendered the victim semistuporous and liable to be careless with fire. Once the individual's clothes had been set well alight, the body fat melted, fueling a slow and particularly complete combustion of the entire corpse. Dupuytren's conclusions agree well with the modern explanation of the "candle effect."

Another disbeliever was Jöns Jacob Berzelius, one of the founders of modern chemistry. He was not at all impressed with the flood of second-rate research on spontaneous combustion emanating from the German and French universities. In a commentary presented to the Swedish Academy of Sciences

6   he declared himself skeptical about "these miraculous happenings" and concluded that "it would be advisable not to construct theories to explain this phenomenon until its existence has been irrefutably proven by unbiased scientists." Many a later "combustionist" would have benefited by study of the great chemist's verdict.

In nineteenth-century Britain, spontaneous human combustion was accepted as a reality. Novels, newspapers, and magazines continually emphasized the danger of bursting into flames during a period of intemperance, and the medical journals regularly published case reports. The *London Medical and Surgical Journal* in 1832 published the astonishing tale told by Dr. C. L. Devouard of Titchfield Street, London. A sailor and part-time smuggler named Thomas Williams, who was in the habit of consuming more than two quarts of strong rum daily, one day took substantially more than that amount. His friends carried him home and called a surgeon, but when they held a lighted candle near Williams's body, the flesh of the wretched man ignited and burned with a blue flame. All efforts to extinguish the fire failed, and finally, the entire body, except the head, legs, and parts of the arms, was consumed, though Williams's flannel shirt was not burned. His friends kept it as a curiosity. They believed that the Devil had set the sottish mariner alight, bringing him down to the shades below for his wickedness.

Another remarkable case, which was perhaps more truthfully described, occurred near Nairn in Scotland. A carter of wood named John Anderson had drunk, for several years, a common bottleful of ardent spirits each day in addition to beer and porter. Nevertheless, he was said to drive his cart and manage his horses tolerably well, and he liked to smoke his pipe as he was driving homewards. One day smoke could be seen rising from the cart. Anderson had caught fire. He got out of the cart, staggered, and fell; when a herdboy called for assistance fifteen minutes later, the body was black, disfigured, and burning. John Grigor, M.D., of Nairn examined the body and stated as his opinion that it was no ordinary combustion but "a case of progressive igneous decomposition" due to alcoholic excess. The editor of the *London and Edinburgh Monthly Journal of Medical Science*, in which the report was published in 1852, doubted Dr. Grigor's theory.

‽ The early temperance movement took the spontaneous combustion craze as a God-given opportunity to have a serious talk with the innumerable alcoholics of the early nineteenth century. Even the most hardened topers, who shrugged their shoulders at grisly descriptions of alcohol-induced dementia and insanity, and laughed drunkenly when the drawn-out torments of hepatic failure were described, sat pale and trembling before the threat of the blue flame of intemperance, taking it as a foretaste of the hellish torments awaiting them below. An essay on spontaneous combustion by the Frenchman Pierre Aimé Lair, written in 1799, was particularly popular among the temperance propagandists because even for the subject, it was unusually rich in macabre detail. The spontaneous combustion myth died hard among fanatical teetotalers, and even late nineteenth-century temperance publications include horrific accounts of the fiery death in all its variety. Some drunkards belched flames like a factory chimney and died after the most miserable agonies; others met their fiery fate when a flammable belch or fart caught fire at a candle, and the flame resisted every attempt to extinguish it. Under the headline "Fire! Fire! Blood on fire!" the February 1863 issue of the *Pennsylvania Temperance Recorder* told the story of a drunkard who submitted to bloodletting; a pint bowl of his blood was ignited, and it burned with a vivid blue flame, to the horror of the wretched patient. As late as 1913 a French temperance magazine published a story about a dipsomaniacal omnibus conductor who smoked a cigarette while clipping the tickets; he ignited spontaneously and burned like a torch in front of the startled passengers. In 1905 when the Norwegian physician Johan Scharffenberg wrote an article in a temperance journal ridiculing the old fad of spontaneous combustion, many readers wrote angry letters to the editor, insisting on the reality of the phenomenon.

In Germany the common people had great faith in liquid manure as a preventive for spontaneous combustion; when a simple-minded farmhand or a village idiot believed himself at risk to burst into spontaneous flame after a good booze-up, he shoveled large quantities of liquid manure and muddy water into his mouth. It is telling that this preposterous cure also appeared in Dr. Carl Rösch's serious scientific work on alcohol abuse, *Der Missbrauch geistiger Getränke* (1838). Scandinavian folk medicine recommended an even more unlikely substance to extinguish the spontaneous flames. Swedish and Norwegian folklore prescribed human urine, preferably freshly voided by a woman, to be thrown into the flaming mouth of the burning drunkard. One laconic nineteenth-century case report from Västergötland (Westgothia) describes the fiery fate of a habitual toper named Zakris, who suddenly ignited spontaneously: "One day, Zakris of Hester burned with a blue flame while he lay in bed. His wife sat astride his head and put out the flames by pissing in his mouth; only a woman could do this. After this, he did not drink aquavit any more."

These tales of drunkards igniting spontaneously and burning with a blue

8   flame issuing from their mouths appeared very early in Scandinavian folk tradition. What appears to be the first recorded case of spontaneous human combustion dates from the early sixteenth century, but was fatal only to the attending physician. Gampe-Saevrei, parson of the village of Rauland in Telemark, Norway, was leaving church after Sunday service. He saw a man lying dead-drunk on the church green, the blue flame of spontaneous combustion blazing from his open mouth. The philanthropic parson, who wanted

Fig. 1. A greedy old German money changer combusts spontaneously as a divine punishment in this broadsheet from 1621. From the author's collection.

to save the life even of this intemperate nonchurchgoer, "pissed on the man to extinguish the flame." The patient was rudely awakened from his drunken stupor by this treatment, and since he knew nothing of the current therapy for spontaneous combustion, he was highly affronted. The congregation took his part, and Gampe-Saevrei had to run for his life. He managed to reach a rowboat, but the howling mob followed him in other boats. After a furious chase on land and sea, he was beaten to death with the candlestick from his own altar. A sacred well was said to appear at the site of the priest-murder, and it was shown by the devout villagers for many years.

❧ Well into the mid-1800s, the belief in spontaneous human combustion was widespread throughout Europe. Although some chemists doubted the existence of this phenomenon, the majority of medical men and forensic experts still accepted it. The sinister death of a German lady would change many minds. On June 13, 1847, the charred corpse of the countess of Görlitz, a noble lady of temperate and religious habits, was found in her bedroom. The municipal medical officer, Dr. Johan Adam Graff, was called in to examine her remains. The corpse was lying on the floor near a partially burned writing desk, the head toward the windows of the room. Although the face was badly charred, Graff could see that the mouth was wide open and the tongue protruding. Another pathologist might have concluded that the countess had been strangled, but Dr. Graff was a "combustionist," and he pronounced the death another case of spontaneous incineration. To be sure, the circumstances were unusual, for the poor countess had been a middle-aged lady of healthy and vigorous constitution, and her consumption of wine — half a bottle a day — did not seem in any way excessive. The police made no further investigation, and the corpse was buried without autopsy.

The strange happenings at the von Görlitz mansion attracted much curiosity among the Darmstadt gossips. Some unscrupulous German newspapermen did not hesitate to declare that the countess had been murdered; some accused Johann Stauff, a valet with a bad reputation, and others blamed the count von Görlitz himself. In order to clear his name, the old nobleman had to demand a further inquest into his wife's death. Soon after, the count's cook found verdigris in a sauce intended for his lordship's table; this dish had been carefully prepared by the valet Stauff. Apparently, the audacious menial had attempted to murder the count to prevent an inquest from being held, and the police took him into custody. The inexorable German legal machinery proceeded through three years of trials, appeals, medical certificates, and exhumations of the body. Stauff was further incriminated

10   after his father was arrested while trying to sell a large lump of melted gold, whose origin he could not satisfactorily explain; it was most likely the remains of the countess's jewelry.

When Johann Stauff finally faced a murder trial, his attorneys vigorously maintained that the countess had combusted spontaneously, but they had great difficulty explaining a fissure in the corpse's skull. The prosecutor claimed it was the result of a blow. The court president called in two eminent scientists from the nearby university of Giessen, the chemist Justus von Liebig and the anatomist Theodor Ludwig Bischoff, to serve as expert witnesses. Bischoff got hold of a corpse and various combustible substances and tried to reconstruct the fire in the cellars of the Giessen Institute of Anatomy. He discovered that a large quantity of external fueling was required for a combustion as complete as that of the countess's body and that the fissure in the skull could not have been caused by the heat, and he handed around the head of his experimental corpse to prove it. Justus von Liebig claimed that the high water content of the human body would prevent it from combusting spontaneously, whether or not the tissues contained a large quantity of alcohol or the individual was excessively fat. The scientific testimony apparently convinced the jurymen, and they convicted Johann Stauff of murder, robbery, and arson; he was sentenced to life imprisonment.

From an early age, Stauff had been known as a slippery, dishonest character who could not live within his income; many people had suggested that his talents would be better employed in America, but he lacked the price of passage. He later confessed that, after the countess had caught him red-handed stealing money from a cupboard, he had knocked her down and then strangled her. He put the corpse on a chair and stacked combustible substances near it, setting it alight to hide his crime. It appears that Stauff was little more than a common ruffian, and he certainly had no knowledge of the lore of spontaneous human combustion; it was just chance that made him burn the corpse after he had strangled the countess, thereby nearly getting away with murder.

The public relished the macabre and exciting details of the von Görlitz trial; newspapers all over the world reported its outcome, and the infamous German servant soon became known as the "demon valet." Both Bischoff and Liebig published the results of their investigations as pamphlets, and many medical journals published extracts from them, along with extracts from the trial. This led to a swift change in opinion among medical men throughout Europe: before the convincing evidence in the von Görlitz case, most people agreed that spontaneous human combustion was merely an old

fallacy. The combustionists rallied around a long article by the French forensic scientist A. Devergie, in which Liebig's theories were criticized, but their
number decreased swiftly during the 1850s. One ingenious argument from
the opposition was that according to the common laws of physiology, not
only human beings but also animals should belong to the population at risk.
There were no reports of spontaneously igniting cows and sheep lighting up
the German landscape at night like beacons or of faithful Fido, sleeping near
the fireplace, suddenly being transformed into a *hot dog*. When, in 1860, the
influential German forensic scientist J. L. Casper took Liebig's part in his
famous *Praktisches Handbuch der gerichtligen Medicin*, declaring spontaneous
human combustion to be mere superstition, it was the final blow for the combustionists in Germany. From the year his book was published to the present
day, not one word has been written on spontaneous combustion in the German medical journals.

❧ Although spontaneous combustion had been largely discredited on the
European continent, the theory died hard in Britain and the United States.
In 1870 the well-known British physician Alexander Ogston published an
extensive review on this subject, expressing serious doubt whether the old
legend of the fiery death could really be completely disregarded. He was
particularly impressed with the high percentage of alcoholics among its victims, suggesting that some obscure process might render their tissues flammable. His father claimed to have met with a case in which the body was so
saturated with alcohol that the fluid in the ventricles of the brain caught fire
and burned when lit with a candle. The cerebrospinal fluid of another individual, who had died after drinking a quart of gin, resembled gin to a marked
degree, "as tested by the smell, taste and inflammability." In 1888 Dr. J.
Mackenzie Booth, of the Aberdeen General Dispensary, was called to examine the remains of a sixty-five-year-old pensioner of notoriously intemperate habits. The body was almost a cinder; yet it had kept the form and
figure of the old soldier, including his military mustache. Three years later,
Dr. Ernest S. Reynolds, physician to the Ancoats Hospital, saw a similar
case in Manchester. Like Ogston and Booth, he denied the existence of true
spontaneous combustion, but he believed that chronic alcoholism might
sometimes lead to the deposit of flammable substances in the tissues of the
body, which explained these remarkable cases of increased combustibility, in
which the individual's body was found almost totally consumed by fire, while
flammable objects nearby escaped the flames. A friend of his, Superintendent
Tozer of the Manchester Fire Brigade, volunteered the information that he

had several times seen human bodies burn vividly without additional fuel and that he was sure that old alcoholics would burn much more easily than other people. As late as 1905, several medical men objected when the *British Medical Journal* mocked the old fallacy of spontaneous human combustion, sending letters and case reports to the editor.

In the United States belief in spontaneous human combustion persisted even longer. In the *Therapeutic Gazette* of 1889, Dr. Archie Stockwell published a lengthy review detailing many cases, both foreign and American. Some months later, another medical man described the simultaneous spontaneous combustion of an elderly married couple at Seneca, Illinois.

Three years later an even more remarkable case appeared in the *Boston Medical and Surgical Journal*. Dr. B. H. Hartwell was on a call to the outskirts of Boston when he was summoned by a distraught man, who called out that his mother had been "burned alive." Hastily driving to the place indicated, Hartwell found the body of a woman burning at the shoulders, both sides of the abdomen, and both legs. The flames reached fifteen inches above the level of the body, and the clothing was nearly totally consumed. It was evident the body burned of itself, without any external fuel; in fact, the ground was damp after a rain. Her friends said that the victim was fifty-nine years old, of strictly temperate habits. Hartwell presumed that when she set fire to a pile of roots, she had accidentally ignited her own clothes. The case seemed clear evidence to Dr. Hartwell of increased combustibility of the human body.

Fig. 2. The corpse of the soldier described by Dr. Booth in the *British Medical Journal* of 1888. From the author's collection.

In the last theoretical article on spontaneous combustion, published in the *New Orleans Medical Journal* in 1893, Dr. Adrian Hava proved a worthy follower in the tradition of far-fetched hypotheses on this subject. Certain that spontaneous combustion existed, he performed a series of bizarre experiments to prove his point. First, he thrice daily administered a dram of brandy to twelve young roosters, no doubt hoping that one or more of them would some day spontaneously become a roasted broiler. His hope was not rewarded, though he continued the experiment for fourteen months, until only two roosters remained alive. The doctor then came up with another odd notion: that the alcohol was being metabolized into carbon monoxide, which was retained in the tissues, making them flammable. He devised another cruel experiment, putting roosters, rabbits, pigs, and other animals in a large glass case filled with carbon monoxide. The wretched animals were not taken out until they were unconscious. After several years of these bizarre experiments Dr. Hava claimed that a rabbit could be set alight after "169 days of continued and careful administration" of carbon monoxide, whereas a rooster took over eight months.

In twentieth-century medical literature, nothing much is said about spontaneous human combustion. There were no more grisly case reports, and fortunately for the animal kingdom, no one tried to replicate Dr. Hava's experimental studies. There is, however, one quite recent case of "spontaneous combustion" on record, in the *Annals of Emergency Medicine* of 1991. A thirty-two-year-old habitual drunkard was taken into the emergency department of Boston City Hospital for evaluation of a seizure. He had another seizure while he was there, and his attendants were mystified to see that the seizure activity changed after a few minutes, the man arching his back and frantically slapping his thighs. They were even more surprised when smoke billowed up from around his abdomen, and he appeared to be spontaneously combusting. Once his clothes were removed the doctors discovered the cause: two books of matches in his back pocket had rubbed together during the seizure until they burst into flame. Apparently, the patient suffered no permanent ill effects from his unique experience.

By the mid-1900s the medical community had forgotten spontaneous human combustion, and historians mentioned it only to scoff at a ridiculous remnant of the age of superstition. But the fiery death received new supporters among the devotees of paranormal phenomena. That well-known researcher of strange happenings, Charles Fort, was the first to assign a supernatural cause to this phenomenon, ascribing it to the work of "things or

14❧ beings." Like the werewolves, of whose existence Fort had as little doubt, they preferred female victims. Several other theorists of the occult firmly supported the notion of spontaneous human combustion, none more than the British writer Michael Harrison, who published the book *Fire from Heaven, or How Safe Are You from Burning?* in 1976. Its preface declared it to be the definitive monograph on its subject and described the evidence that people might, indeed, combust spontaneously as irrefutable. The book attempts to link spontaneous human combustion with psi phenomena and poltergeist activity, and it draws parallels to the destruction of Sodom and Gomorra. In all earnestness, it suggests that the combusting individuals were taken to "Some Other Place, for examination by Some Other People," but it does not explain why presumably superintelligent aliens would prefer to converse with old alcoholics, from whom little reliable information could be gleaned. Another imaginative "occult philosopher" has applied Michael Harrison's theories to come up with an even more startling notion, published in the journal *Medical Hypotheses* in 1983: the Resurrection of Christ could easily be explained if the body was completely destroyed by spontaneous combustion, which would also explain the scorch marks on the Shroud of Turin!

Michael Harrison's book has been by no means the only one devoted to the "SHC" phenomenon, as it is known among the initiates. Indeed, considering the flood of amazing new theories, *Fire from Heaven* might seem a rather staid and conservative "standard work." The great Jöns Jacob Berzelius, who warned against undue theorizing until the existence of spontaneous human combustion had been established with certainty, would have been aghast at the harebrained speculations on this subject by latter-day amateur scientists. Some still favor the poltergeist hypothesis; others blame "man's electrodynamic being" or geomagnetic field fluctuations; in his book *Le feu qui tue*, published in 1986, the French writer Emmanuel Peze attempts to link spontaneous human combustion with UFO activity. Others suspect the "Kundelini fire," an energy that is supposed to pervade the individual's astral body, according to the teachings of Yoga. Chemical theories abound, some even more imaginative than those of the nineteenth-century German thesis writers. SHC is attributed, for example, to an excessive accumulation of phosphoenolpyruvate, a particularly "energy-rich" compound. Nor have nineteenth-century theories died. There are still those who believe that excessive use of alcohol may cause deposition of flammable, nitroglycerinlike phosphagens in the muscle tissue, or that a buildup of electrostatic charge within the human cells can split the body water into oxygen and hydrogen.

The writer Larry Arnold turned to physics, postulating the existence of a "pyrotron," a hitherto unknown subatomic particle, capable of starting a chain reaction that vaporizes the entire human being.

Today, belief in spontaneous human combustion seems to be steadily increasing among the general population. Popular works such as the Reader's Digest's *Mysteries of the Unexplained* and others of that ilk pander to the public obsession with "unsolved mysteries" with grisly tales of spontaneous combustion, along with a veritable hodgepodge of ghost stories, UFO close encounters, and the horrors of the Bermuda Triangle. Furthermore, the tabloid newspapers, both in Britain and in the United States, are firm supporters. Some recent headlines would have delighted a nineteenth-century temperance writer: "Preacher Explodes into Flames in the Pulpit," "Demons Burned My Hubby to a Crisp," and "Man Eating Chili Bursts into Flames and Dies"!

From 1600 to 1900, ninety-seven cases of spontaneous human combustion or increased combustibility were reported in the medical literature of western Europe and the United States. A considerable proportion (34 percent) occurred in France, 20 percent in the German states, 16 percent in Britain, and 16 percent in the United States. A majority (about 70 percent) of the victims were women. Remarkably, in not less than 65 percent of all cases, it was positively stated that the victims were alcoholics, often with a history of many years of abuse. Michael Harrison's book and the more recent *Spontaneous Human Combustion* by Jenny Randles and Peter Hough add a good many mysterious twentieth-century fire incidents of varying reliability. The scientific world has greeted these silly and credulous books with mere ridicule, but were the hard facts extracted from them and all the wild theories and unreliable newspaper accounts weeded out, the result would be a slim scientific paper presenting at least twenty-five verified twentieth-century instances of preternatural combustibility.

Modern forensic science's policy with regard to spontaneous human combustion has largely been that of the ostrich: only two scholarly papers on increased combustibility are on record in this century. One of these was published by Dr. Gavin Thurston in 1965. A slim widow of abstemious habits was found dead one morning, her charred body lying in the hearth. Her legs, including stockings and slippers, were quite intact, and there was little damage to the surrounding furniture, though the room was full of soot and cinders, just as in the classical cases. It was evident from the postmortem that she had advanced cardiovascular disease, which was probably the cause of death; it was further presumed that she had fallen into the fireplace after death and that the body had burned in its own fat, with the clothes acting as a wick — a phenomenon sometimes called "the candle effect."

The most famous of all modern cases of alleged spontaneous combustion took place in 1951 in St. Petersburg, Florida. Mary Reeser, an obese sixty-seven-year-old widow, had last been seen sitting in her armchair, wearing a flammable nightdress and dressing gown. She had taken sleeping pills and was a smoker. The next morning, the landlady found the doorknob of Mrs. Reeser's apartment too hot to grasp. The room was a horrible sight, full of greasy soot "with a peculiar odor." Of the armchair only the coil springs were left; of Mrs. Reeser only a heap of ashes remained, together with a small roundish object identified by some as the shrunken head, some lumps of charred tissue, and the left foot, with the skin unburned and a black satin bedroom slipper still in place. Newspapers on a table nearby were not even scorched, but two candles had melted and a clock had stopped at 4:20.

Several distinguished forensic scientists and fire analysts, as well as the experts of the FBI, had to declare themselves baffled by this strange case, which resembles the classical spontaneous combustions in many particulars. A close reexamination of the Reeser case, performed by the American forensic analysts Joe Nickell and John F. Fisher in the late 1980s, has cleared up many of the obscure points. They have proposed that Mrs. Reeser set herself alight with her cigarette while sitting in her armchair in a stupor induced by the sleeping pills. Her body fat melted into the stuffing of the chair, leading to a very complete combustion of the body. That one foot did not burn was probably explained by the fact that she had a stiff leg and the foot fell outside the radius of the fire. The globular object found at the scene was probably the neck, not a shrunken head, for the intense heat must have burst the skull. The extensive destruction of the body they ascribed to the "candle effect," here facilitated by the chair stuffing, which acted as an enormous wick, far more effective than the victim's clothes alone would have been.

Such a "prolonged human combustion" takes considerable time: the body fluids gasify and the melted fat burns with sufficient heat to destroy bone and internal organs. In a later, even more puzzling case, described in the *New Scientist* in 1988 by John Heymer, a seventy-three-year-old man combusted very completely, sitting in his armchair before the fireplace. Only a heap of ashes, a blackened skull, and both legs, with socks and parts of the trouser legs, remained. It was presumed that he had slumped forward into the fire, thus igniting his clothing, and then fallen back into his chair, where his body had been burned even more completely than Mrs. Reeser's.

From the earliest literature, it is evident that the notion of spontaneous human combustion emanated from the sixteenth- and seventeenth-century popular belief that the drinking of strong spirits might light a spontaneous flame in the stomach. Since this belief had its origin in Scandinavia, it was no coincidence that the Dane Thomas Bartholin was the first scientist to assimilate it into seventeenth-century medicine. The idea of aquavit drinkers bursting into spontaneous flames thus preluded the first mentioning of the phenomenon we can call increased combustibility by more than 100 years; the first of these odd postmortem reports of bodies found extensively destroyed by fire without major damage to the surroundings was published in 1673. Many others have followed, and at least 120 well-attested cases of this phenomenon exist on record. There has been no satisfactory instance of any individual combusting *spontaneously*, and in the greater part of the cases an external source of fire is apparent.

The "candle effect" explanation put forth by modern forensic experts seems perfectly applicable to the majority of both new and old instances of preternatural combustibility, but the lack of any solid experimental data of this phenomenon leaves some important questions unanswered, mainly whether some people are more combustible than others, as many writers have presumed. One would imagine that the bodies of very obese people were more susceptible to undergoing combustion by the candle effect, but many victims have been slim. Dehydration and old age have been proposed as risk factors by some writers, but without much hard evidence. The preponderance of alcoholics among historical victims, although not among twentieth-century ones, remains a mystery. It is very unlikely that their bodies were really more combustible. Instead, a drunken individual is prone to be more clumsy with fire, and might be too befuddled, drowsy, or even unconscious to escape. Furthermore, some early accounts would imply that clergymen and temperance advocates sometimes exaggerated the drinking habits of the victims of increased combustibility. It is likely that external factors were considerably more critical. The quality of the individual's clothing ought to have great importance: it had to be flammable, yet burn slowly, generating enough heat to crack the skin and melt the subcutaneous fat, as well as being durable enough to act as a wick during the combustion process. The presence or absence of a draft in the room may also have been important. In theory it would be easy to test these hypotheses, but ethical problems would naturally arise about the procurement of suitable corpses for experiment; the time-honored device of snatching fresh corpses from the morgue or churchyard has long been discredited. Nor would the university grants committee encourage a researcher who proposed to drop lighted pipes on two hundred unconscious pigs sitting in easy chairs and wearing human clothing ranging from silk night dresses to jackets and plus fours, some in drafty rooms and others in sealed ones.

❧ Captain Marryat's once famous novel *Jacob Faithful,* which was first published in 1834, is not frequently read today; it has been labeled a "modern classic" and vainly struggles against literature with a greater capacity for fascinating the youthful mind. Perhaps its ability to compete with modern pulp novels would have been increased if the present abridged edition had not excluded the powerful scene in the first chapter, depicting the death of Mrs. Faithful, Jacob's mother, from spontaneous combustion. This lady was very fat, and her son used to say that "locomotion was not her taste — gin was." She usually sat all day in her cabin, below decks in the family barge. One evening when Jacob was on deck and his parents sat drinking below, he heard a terrible scream. His father came running up the hatchway and dived into the river without a glance at his son, never to be heard of again. Clouds of thick, empyreumatic smoke billowed from the hatches, and poor Jacob could not run down to see what had happened to his mother. When the smoke had cleared ten minutes later, the unhappy lad saw that he had lost both his parents at a single stroke, one dead by water and the other by fire: only an unctuous pitchy cinder in the middle of the bed was left of his mother's voluminous body, and his father had gone mad when she combusted spontaneously before his eyes. The bed with her remains was taken ashore for further investigation of her strange death, but only one conclusion was possible. Poor Mrs. Faithful was a textbook case. Thousands of people flocked to see her grisly remains, and Jacob's guardian collected money for the young orphan from them, before selling the bed, cinder and all, to a doctor for twenty guineas.

Captain Marryat was not the first writer to use spontaneous human combustion in a novel. The phenomenon proved useful as a dramatic, horrifying climax and a convenient means to dispose of a "spent" character. Already in 1798 the American Charles Brockden Brown had had an old man erupt into flames in the novel *Wieland,* as a divine judgment; this event plunges his son into a bout of melancholia, and he ends up butchering his entire family. In the novel *Redburn*, published in 1849, Herman Melville followed Marryat in having the combustion occur aboard ship. When the sailors notice a strange smell in their cramped quarters, some attribute it to a dead rat, but others take a look at their dead-drunk shipmate lying in his bunk:

> "Blast that rat!" cried the Greenlander.
>
> "He's blasted already", said Jackson. . . . "It's a water rat,
> shipmates, that's dead, and here he is."

As they speak, flames flicker over the man's face. Later, the burning body is lowered into the sea, where it sinks "like a phosphorescent shark." Some other American writers, including Mark Twain and Washington Irving, also had minor characters combust spontaneously, in order to enliven the narrative. Nikolai Gogol in his famous *Dead Souls* had a drunken blacksmith die enveloped in the blue flames of intemperance. Another remarkable literary spontaneous combustion occurs in *Un capitaine de quinze ans* by Jules Verne, set in Angola in early 1873. A native ruler, the king of Kazonnde, falls victim to his craving for the white traders' spirits and ignites spontaneously while drinking a bowl of flaming punch. The king burns like a demijohn of paraffin, and when one of his ministers tries to save him by smothering the fire, he too bursts into spontaneous fire. If their ambition had been to "strike a light in darkest Africa," the king and his chancellor succeeded beyond their wildest dreams.

Jules Verne was known for his imagination, famous as a progenitor of science fiction, and therefore he might be expected to include such an incident in one of his novels. It is more remarkable that the great naturalist Emile Zola depicted a similar incident in the novel *Le Docteur Pascal*, published in 1893 and translated into English by Mr. Ernest A. Vizetelly the same year. When Félicité visits her brother-in-law, the drunken old M. Macquart, she finds him sleeping in his chair, a lighted pipe lying in his lap. Through a hole in his trousers, she can see his thigh burning with a blue flame. "Now the liquid fat was dribbling through the cracks in his skin, feeding the flame which was spreading to his belly. And Félicité realized that he was burning up, like a sponge soaked in alcohol. He had been saturating himself for years with the strongest, most inflammable of spirits. Soon, doubtless, he would be flaming from head to feet." Zola's revival of spontaneous combustion lore caused some consternation among the contemporary critics, and although Zola himself denied it, the reviewers presumed that he had been influenced by the earlier literary spontaneous combustions. The main character of the novel, a skillful physician, is able to make a spot diagnosis when he later has

Fig. 3. The sad end of the king of Kazonnde, an illustration by Henri Meyer for the 1879 English translation of Jules Verne's *Un capitaine de quinze ans*. Reproduced by permission of the British Library, London.

occasion to see the remains of poor M. Macquart. With Gallic eloquence, Doctor Pascal discusses this strange occurrence:

> "Now he died a royal death, like the prince of drunkards, blaz-
> ing away of his own accord, consumed in the funeral pile of his
> own body!" Carried away by his own enthusiasm, the doctor
> flung out his arms, as if to embrace the imaginary scene: "Can
> you see it? So drunk that he could not feel himself burning,
> lighting himself like a Saint-Jean's fire, dispersing himself in
> smoke, first spread over the four corners of this kitchen, dis-
> solved in and floating in the air, bathing all objects he owned,
> then escaping in a cloud of vapour through this window as soon
> as I opened it; flying off into the sky, filling the horizon? . . .
> Verily an admirable death: to disappear, to leave nothing of one-
> self behind, except a handful of ashes and a pipe by one's side!"

A little-known literary spontaneous combustion occurs in Edwin Brett's novel *The Darling of Our Crew*, a story for boys which appeared in 1880. On board a ship, the purehearted young hero of the story looks on aghast while a witchlike hag called Mother Shebear seizes a pint of raw brandy and puts it over the fire until it seems on the point of exploding. She blows out the blue flames visible at the top and drinks half of it at a draft, chucking fiend-ishly. Advancing upon the petrified lad, Mother Shebear stretches forth her long talons and shrieks, "Now, boy, you shall be my victim. Ha! Ha! Ha!" But before she can toast poor Jack over the fire, Mother Shebear is herself ignited by the brandy. A bluish lambent flame bursts from her mouth and curls in fantastic wreaths about her face. The boy tries to extinguish the flames with water, but it turns at once to steam. Before his eyes, the hag becomes "an incandescent mass of matter like a statue, burning and glowing like the center of a fire."

A no less ludicrous literary spontaneous combustion occurs in *Husmanns-gutten* (The farm laborer's son) by Hans Andersen Foss, a late nineteenth-century Norwegian author of very popular rural melodramas. The novel's young proletarian hero is prevented from marrying his fiancée by her father, a haughty country squire much given to drunkenness. The disappointed young Norwegian then emigrates to the United States, where he works at a factory for two years, saving every penny he earns before returning to his native country. With his newly acquired fortune, he purchases the impover-ished squire's house and farm and settles there with his wife, the squire's daughter. While traveling home from a drinking party in his horse-drawn sleigh, the squire ignites spontaneously. Blue smoke belches from his innards up through his open mouth — a strange spectacle indeed for any passersby in the cold, clear Scandinavian night. This absurd old novel was one of Norway's best-selling books of all times and was even filmed in the 1930s.

No less a poet than Johann Wolfgang von Goethe was inspired by the fiery death in his *West-östlicher Divan*, which was written in 1819, when the spontaneous combustion craze reigned in Germany:

> *Schenke hier! Noch eine Flasche!*
> *Diese Becher brign ich ihr*
> *Findet sie ein Häufschen Asche,*
> *Sagt sie: Der verbrannte mir.*

I translate the verse in the following way:

> *Another beaker! How it flashes!*
> *This I consecrate to thee;*
> *If you find a heap of ashes,*
> *Say, he was consumed by me.*

One of Sweden's finest poets of the late nineteenth century, Gustaf Fröding, also thought about spontaneous combustion, as evidenced by one of the poems in the collection *Stänk och Flikar*, presented here in my translation:

> *I am a bed of blackened coal,*
> *Of burned-out lust and death;*
> *I breathe a flame with alcohol,*
> *The fire is on my breath.*

☙ The best known of all literary spontaneous combustions is that of "Lord Chancellor" Krook in the Charles Dickens's famous *Bleak House*. Dickens's fascination with spontaneous combustion is already evident in his early work. He mentions it in describing a fainting woman in the article "The Prisoner's Van" written in 1835: "All the female inhabitants run out of their houses, and discharge large jugs of water over the patient, as if she were dying of spontaneous combustion, and wanted putting out." In *A Christmas Carol*, when the room is suddenly illuminated by the coming of the second spirit, Ebenezer Scrooge believes that he "might be at that very moment an interesting case of spontaneous combustion, without having the consolation of knowing it." In *Bleak House*, Mr. Krook is a cadaverous, withered old rag-and-bone seller, whose breath is "issuing in visible smoke from his mouth, as if he were on fire within." One day, when the gin-sodden old man is sleeping in front of the fireplace, he ignites spontaneously, and foul-smelling, greasy smoke billows out through the windows. Some people walking by presume that someone has burned his chops on the stove, and that they were not particularly fresh to begin with. When Mr. Snagsby and Mr. Guppy enter Krook's rooms, they find them full of a greasy soot, the windowsill covered with a repulsive yellow fluid. They suspect that Krook has committed suicide and try to find the body. Finally, they see Krook's vicious cat, Lady Jane, stand snarling before his empty chair, on the seat of which a small piece of coal can still be seen. They realize what has happened, and with the cry "Horrors! He *is* here!" they rush out into the street.

It is apparent that Dickens based Krook's death on one of the most famous eighteenth-century cases, that of Countess Bandi in 1731; many details, such as the soot and the yellow fluid on the windowsill, were given almost verbatim. It is unlikely that Dickens had studied the original literature, but the lore of spontaneous combustion was extremely widespread in his time, and the Italian countess's fiery fate was retold in many books and periodicals. Dickens could easily have read of it in either the *Gentleman's Magazine* or the *Annual Register*, and one literary historian has suggested that he also studied the section on spontaneous combustion in Robert Macnish's *Anatomy of Drunkenness*.

Fig. 4. "The Appointed Time" from *Bleak House*. A drawing by "Phiz" in the 1933 edition of Dickens's collected works. From the author's collection.

24   Other writers have implied that Dickens found his inspiration in the *Terrific Register*. This loathsome periodical abounded in bloodcurdling tales of murder, rape, incest, torture, and dismemberment, and its editors did not sneer at detailed accounts of spontaneous combustion, including a grisly description of the Italian countess being reduced to ashes. The *Terrific Register* had a considerable readership at its time, including, as he himself later confessed, the young Charles Dickens, while he was at Wellington House Academy.

     Charles Dickens was one of the leading novelists of his time, and many thousands of people eagerly read each new installment of *Bleak House*. One of them was the literary critic George Henry Lewes, who was *à jour* with the scientific debate on the spontaneous combustion question, including the findings of Liebig and Bischoff. He publicly faulted Dickens for reviving a vulgar error, and Dickens took umbrage. He evidently believed that spontaneous combustion was possible, and instead of claiming artistic license, he attempted to prove that this phenomenon really existed. Since he was out of

Fig. 5. A drawing inspired by the spontaneous combustion controversy between Dickens and Lewes, from *Diogenes*, January 8, 1853. From the author's collection.

ALARMING CASE OF SPONTANEOUS COMBUSTION.

" Oh ! Law ! there's. Pa's boots—but where's Pa ?"

his depth in medical matters, he turned to his friend Professor John Elliotson for support. Unfortunately, the professor was a rather credulous character and not the soundest guide in these matters. He was a zealous phrenologist, and his stubborn defense of Mesmer's teachings, by which he claimed to cure everything from headache to cancer, had cost him his chair at the London University College. In spite of Elliotson's dubious reputation, Dickens relied wholly on a manuscript of his, quoting many older sources from the medical literature. Lewes and his literary associates were unimpressed by these arguments, and Dickens was generally considered the loser of the debate; he was even ridiculed in the comic magazines as the last champion of spontaneous human combustion. Nonetheless, even his critics had to agree that the section on Krook's death had the master's touch; it is today frequently quoted as an example of Dickens's power to create an atmosphere of impending horror. Dickens remained unrepentant, and when *Bleak House* was published as a book in 1852, he vigorously defended the fiery death in the preface, and the chapter about Krook's death ended with the same fateful words: "It is the same death eternally, engendered in the corrupted humours of the vicious body itself, and that only — Spontaneous Combustion, and none other of the deaths that can be died."

For several years after *Bleak House* was published, the existence of spontaneous human combustion was debated in the newspapers and in private conversation. According to the *Life* of the sculptor Thomas Woolner, Krook's case came up at a country house where Woolner was staying. When a scientific gentleman made the point that it was very dangerous for a drunken person to go near a candle or any open fire for fear of igniting spontaneously, the Irish butler who was clearing the table was seen to turn pale and hastily leave the room. The host hoped that the talk of spontaneous combustion had scared the butler, who was altogether too fond of drink. Shortly afterward, the butler returned with a Bible, which he placed before a clergyman guest, asking to be sworn immediately. Placing a trembling hand on the Bible, he solemnly pronounced: "I, Patrick O'C. swear *never, never, never to* — blow out a candle again!"

The Bosom Serpent

ACCORDING TO ONE OF THE OLD VIKING
annals, the Flatø Book, King Harald Hårdråde of Norway once visited the nobleman Halldor, whose daughter had been very ill for some time. Fever, increasing abdominal girth, and an unquenchable thirst were the major symptoms. The old women gossiped about her being pregnant, but the young lady denied this with great vehemence. Since her condition was steadily worsening, the king was also consulted. His diagnosis was that she had accidentally swallowed the spawn of a serpent when she drank water, and the reptile had grown within her stomach and nearly killed her. To rid herself of this bosom serpent, she must thirst for several days without being given any water; after that, she was to be taken to a waterfall and there open her mouth wide so that the thirsty snake could hear the streaming water. When it slithered up her gullet and stretched its head out

between the girl's jaws, her father was to strike it with his sword. When the king had left Halldor's house, this cure was performed exactly as he had prescribed. The bosom serpent was decapitated and the girl restored to health. Another Norse legend described the brutal methods used by King Olaf Tryggvason when he was christianizing the heathen Vikings by force. Once, his men were torturing a staunch Icelander, who refused to give up the faith of his fathers. When everything else failed, they put a snake into him. The man tried to save himself by blowing on the snake's head, but the torturers forced it down his throat by applying red-hot irons to its tail. The reptile later thrust its head out through the Icelander's abdomen with the wretched man's heart between its jaws. This dreadful sight, horrific enough for the most exaggerated horror films of the present day, had a powerful effect on the other Icelanders, who eagerly discarded the hammer of Thor in favor of the frail cross of White Christ.

These dramatic stories from the Norse legends illustrate the ancient belief that snakes, frogs, lizards, and other animals can live as parasites within the human gastrointestinal tract. Already in ancient Egyptian, Assyrian, and Babylonian manuscripts, there is mention of a "colic snake" as a cause of painful stomach cramps. In *De morbis vulgaribus*, Hippocrates describes the case of a youth who had drunk a great quantity of strong wine. When he passed out on the ground, a snake slithered down his throat and caused his death from an apoplectic seizure. In his *Contractae ex veteribus medicinae tetrabiblos*, Aëtius lists the many symptoms attributed to the presence of frogs, toads, or salamanders in the human stomach, ranging from fever, vomiting, and stomach pain to tremor, stiffness, and confusion. This diagnosis must often have been considered in difficult cases. In his *Natural History*, Pliny mentions snakes and frogs as gastrointestinal parasites in men and beasts, and the Talmud has several similar tales.

Very early on, however, there seems to have been some opposition to the established belief in the stomach snake. The learned Alexander of Tralles, living in Lykia during the sixth century, was once consulted by a woman who was certain she had a snake in her stomach. He soon understood that she was a hysteric, the snake existing only in her imagination. He asked her to describe exactly what she thought the animal looked like and then procured a similar specimen, which he put in her bowl of expectorations! The woman was completely cured, and she was most grateful to her clever physician, who, one supposes, did not leave the sickroom without pocketing a considerable fee.

28 Snake-Expelling Saints

In medieval times, there was frequent mention of the legend of the bosom serpent, in medical works as well as the annals of the saints. According to a story from the sixth century, a young lad with a snake in his stomach was taken to the holy Monegunde, after many doctors and quacks had pronounced him incurable. With her hand, the holy woman could feel the snake moving in his intestines. She then fastened a small cross to a grape leaf, which was tied to the boy's stomach. This magic talisman had the desired effect: in a vulcanic opening of the bowel, the snake was expelled like a projectile. Similar, albeit less dramatic, tales were told about Saint Simeon and Abbot Hugo de Clancy. One of the votive gifts to Saint Leonhard of Lavantal was a small iron snake, which was given especially by people affected by diseases of the stomach.

The bosom serpent also plays a part in one of the legends about the medical saints Cosmas and Damien, who were martyred in A.D. 283. Their most famous miracle is to have transplanted the leg of a dead Negro onto a church servant who suffered from gangrene. The man was said to have lived for several years with one black leg and one white as a symbol of the devout brothers' operative skills. Another of their miracles concerned a poor peasant who was tortured day and night by a large snake, which had crawled down his throat while he slept. No doctor could help him, but when he stepped

Fig. 1. Saints Cosmas and Damien expel a snake from the stomach of a peasant. From an early fourteenth-century fresco in the Metropolitan Cathedral of Mistra. From the author's collection.

into Cosmas and Damien's church, the serpent slithered up his throat with great haste. Together with the rest of the congregation, the peasant fell on his knees to praise the Lord and the two saints with song and prayers. This miracle is depicted on an early fourteenth-century fresco in the Metropolitan Cathedral in Mistra, as well as on the two medical saints' reliquaries in Munich and Krakow.

The worldwide distribution of the legend of the bosom serpent is well attested in folk tales from many parts of the world. The annals of Buddhadáso mention that in the year 339, an Indian gentleman had the misfortune to get a water snake in the stomach after accidentally having swallowed snake spawn while drinking water. After being tortured by the animal, which gnawed his entrails, he consulted the rajah, who was evidently as skilled as his Norwegian counterpart in curing this disease. The man was well bathed and provided with a comfortable bed, but forcibly kept awake for seven days and nights. When he was finally permitted to fall asleep, he lay snoring with his mouth wide open. The rajah put a piece of meat tied with a string in his mouth. The starved snake crawled up his gullet, bit the meat, and tried to pull it down to its lair. The royal snake-fisherman managed to land his catch, however, and he put it in a bowl of water for the inspection of all and sundry.

In an old Gaelic folk story, a beggar with newts in the stomach was cured, again by a prince. Royalty seems to have been invested with great curative powers in these difficult cases! This time, the man was given salt meat to eat. He then knelt by a stream of water, and the thirsty newts raced up his gullet and jumped into the brook.

In several ancient folk tales from the Balkan Peninsula, important parts are played by snakes living in the human stomach. In one of these tales, a wicked stepmother tricked her naive stepdaughter into drinking water containing snake spawn. As the snake inside her grew, the poor girl's stomach became large and swollen. The stepmother then accused her of being pregnant as the result of an illicit intrigue, and her family ostracized her. She was employed as a goosegirl at a royal farm. When she fell asleep in the fields one day, the snake slithered up her throat and thrust its head out through her mouth, like a periscope, in order to take its bearings. A noble young prince, who was fortunately passing by, cut it in two with his sword. The tale concludes with the customary marriage and happiness for the young couple and frustration of the stepmother's schemes.

A better-known tale is that of King Devasakti, who was taken badly ill after accidentally imbibing a young snake while drinking foul water in a swamp. His evil relatives considered this invalid unfit to reign, and they expelled him from the capital, dressed in rags and equipped with a beggar's staff and begging bowl. After the king, with the snake in his stomach, had wandered from town to town for several years, his luck finally turned. A princess in a distant country had offended her father, the king, in an argument, and as a punishment, he ordered that she had to marry the first criminal or beggar that set foot within the royal castle. This individual was of course the wretched King Devasakti, who came in hope of consulting the court physicians about his stomach ailment. Instead, he found himself forcibly married to the princess and deported to a farm in the provinces. One day, as her husband lay snoring, his head resting on a termite hill, the princess saw a large snake poke its head out of his gaping mouth. The snake conversed with another snake living in the termite nest, and they were unwise enough to detail exactly the only way they could be killed. The eavesdropping princess went home to prepare a terrible laxative for King Devasakti, containing grated sephonantus and mustard, following the blabbering parasite reptile's own recipe: the animal was expelled in the king's copious stools. The enterprising princess then went to the termite hill and scalded the other snake to death with boiling oil and water, again according to its own prescription. She seized the treasure of money and jewels which the snake had been guarding, and with this fortune Devasakti and his wife were able to reclaim their royal positions.

Snakes and Frogs in the Stomach

Many of the collections of curious and abnormal medical cases from the sixteenth and seventeenth centuries contain stories of snakes, frogs, toads, and newts living in the human stomach. Both the *Mirabilia medica* of Marcellus Donatus and the *Observationes medicorum rariorum* of Johannes Schenckius contain several cases of such bizarre parasites. Schenckius was convinced that swallowed frog spawn could adhere to the stomach and develop there, the colony of frogs living and breeding with ease within the human gastrointestinal canal. The frogs would deprive the patient of nourishment. The famous Conradus Gesner reported a strange epidemic that killed more than three thousand people in the city of Theis. Many snakes and newts left their bodies through the mouth or by the stools, and in the stomach of a recently dead young woman, two snakes had been found. Another sad instance of a postmortem diagnosis of a snake in the stomach was reported in the old *Ephemerides*. A shoemaker had committed suicide by stabbing himself after having been tortured by intractable stomach pains for more than ten years.

After burial his coffin was dug up for his wife to have a last look at him. She was horrified to see a snake the length of a man's arm lying beside the corpse. It was presumed that the bosom serpent had made its egress from the body through the stab wound.

In 1618 the prebendary of Strasbourg, Dr. Melchior Sebizius, reported another famous case. A seventeen-year-old youth had consulted him for stomach pains, weakness, malancholia, flatulence, and epileptic seizures, but the doctor was unable to diagnose his ailment. Some weeks later, he was found sitting dead in an outhouse, and beneath the seat, a large snake

Fig. 2. A German broadsheet from 1530, depicting several instances of reptiles as gastrointestinal parasites. To the right can be seen the corpse of a maiden of good family who was found to have two snakes in her intestines at autopsy. From the author's collection.

was crawling about. In a long thesis, illustrated with a figure of the snake, Sebizius concluded that the unfortunate youth's afflictions could all be explained by the presence of the snake in the stomach for an extended period of time, and that the strain of expelling it from his body had obviously caused a fatal apoplectic fit.

In the *Praxis medica admiranda* of Zacutus Lusitanus, living snakes, scorpions, lizards, and hens are spoken of as gastrointestinal parasites, and in the sections concerning live animals within the human body, Lusitanus also describes a woman who gave birth to a salamander and a man who voided a cloud of living flies instead of urine. This medical Münchhausen was considered a somewhat unreliable and imaginative observer even by his contemporaries. Another of his examples was a Capuchin monk who consumed a bowl of salad with too much greed to observe some snake spawn and later voided a long snake by the urethra.

Fig. 3. A drawing of the snake found under the outhouse after allegedly having lived inside the bowels of a young man. From the 1618 thesis by Dr. Melchior Sebizius. The text reads, "A picture of the snake that ruined and destroyed the unhappy youth." From the author's collection.

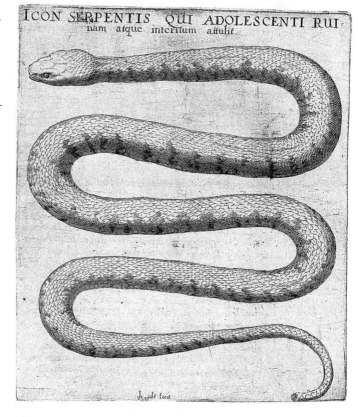

ICON SERPENTIS QUI ADOLESCENTI RUI
nam atque interitum attulit.

The famous French surgeon Ambroise Paré also encountered the bosom serpent, and he tells the story in one of his case reports. In 1561 he met a Parisian woman whom he rather offensively described as "une grosse garce fessue, potelée et en bon point." This corpulent lady of doubtful reputation had caused a great uproar in Paris by claiming that she had a snake in her stomach, which had crawled into her while she slept in a hemp field. Many Parisians came to see her and to feel the snake's motions in her stomach. Ambroise Paré strongly suspected that the whole thing was a fake. He examined the woman with a long speculum, which he introduced into her gullet, but he saw no sign of the snake. When he threatened to give her an extremely powerful laxative, which would certainly get rid her of the parasite, she admitted that the story was pure invention and withdrew from further treatment. Ambroise Paré saw her again six days later, standing near the Montmartre gate and inviting men to come and feel the snake in her stomach!

Two Famous Cases

The most famous of all patients with alleged amphibians in the stomach was Frau Catharina Geisslerin, who was widely known among her contemporaries as "the toad-vomiting woman of Altenburg." In 1642 she took the first step toward notoriety by vomiting several toads and lizards. She claimed that she had accidentally imbibed spawn while drinking foul water in a marsh. The toads, frogs, and lizards were now fully grown and thriving on the chewed food that was delivered to them thrice daily. She could feel them running and sporting in her intestines, especially after she had drunk milk, which the animals particularly liked. After she had vomited toads for two years, the local physician acknowledged that this bizarre case far exceeded his own therapeutic ability, and he invited well-known professors and consultants from all over Germany to come and examine her.

First to accept the challenge was Dr. Thomas Rheinesius, the physician-in-ordinance to the elector of Saxony. His appearance on the scene brought an abrupt cessation of the spectacular toad vomiting: he experimented with various emetics and purgatives for several months but was unable to chivy the animals out of their lair. Professor Michaelis from Leipzig was now called in; Frau Geisslerin was given a powerful emetic, and she vomited something that looked like the leg of an amphibian: the consulting physicians hotly debated whether it belonged to a frog or a toad. In May 1648, after the medical men had left her, Catharina Geisslerin again started vomiting amphibians, producing thirteen toads and a lot of spawn in a couple of weeks. Both Thomas Rheinesius and Professor Michaelis returned to examine the animals further. Rheinesius sent one toad to his old tutor, Professor Thomas Bartholin of Copenhagen, who was particularly interested in this rare case.

When Bartholin dissected the toad, he was astonished to find that its stomach contained more than thirty black, winged insects. Since the animal could hardly have obtained such food if it had spent its entire life inside Frau Geisslerin, the only reasonable explanation would be that the woman had purposely swallowed the toads or that she just pretended to vomit them. Apparently, however, this conclusion did not occur to Bartholin. He later published an article in which he confirmed his belief that toads and lizards could live as parasites in the human stomach. He emphasized the case of the Altenburg woman as one of the most striking ever.

The toad-vomiting woman continued to suffer from her unruly lodgers in the stomach for fourteen more years and threw up specimens at regular intervals. She survived all her medical attendants except Thomas Rheinesius, who visited her as late as 1661 and professed himself greatly astonished that she was still alive. After her death at the Altenburg hospital in 1662, the medical men were of course quite eager to perform an autopsy, expecting to find an interesting vivarium of toads, lizards, and salamanders within her body. To their disappointment, not a single animal was found. The cause of death was given as dropsy and inflammation of the liver.

❧ In 1694 the twelve-year-old son of Pastor Zacharias Döderlein, rector of a Berolzheim village in southern Germany, was taken severely ill. After several apoplectic fits and attacks of abdominal cramps, he vomited a large wood louse. Medical and entomological experts were called in from the faculty at Altorph, and the boy was treated with various insect poisons, but to no avail. During the following three weeks, young Theodorus Döderlein threw up 162 wood lice, 32 caterpillars, 4 millipedes, 2 worms, 2 butterflies, 2 ants, and a beetle. The insects within the young German lad were soon succeeded by amphibians: he vomited 21 newts, 4 frogs, and some toads. These macabre and uncanny happenings soon attracted notice among the clergy. Many German parsons came to visit the house of their stricken colleague, and they concluded, of course, that the boy was possessed by the Devil. It particularly impressed them that when the suffering boy was led to take some fresh air near a pond with croaking frogs, his stomach frogs croaked loudly in reply. When the local doctor was somewhat skeptical about the possibility that a multitude of frogs might be living inside a human being, the clergymen reminded him that if Jonah could have survived within the whale, the reverse should also be possible. They thought it likely that animals could live for a prolonged period of time within the human stomach. The medical men were thus dismissed from the case, and the exorcists took over the treatment of young Theodorus. Their prayers, incantations, and maledictions were initially of no avail. Instead, the boy began vomiting even stranger objects: white and red eggshells, two knife blades, one link from a large chain, two long nails, and a lot of small tacks. The wretched Pastor Döderlein's state of mind can well be imagined when his stern professional brethren demanded an immediate explanation how the Evil One could have gained access to his own household and taken possession of the body of his

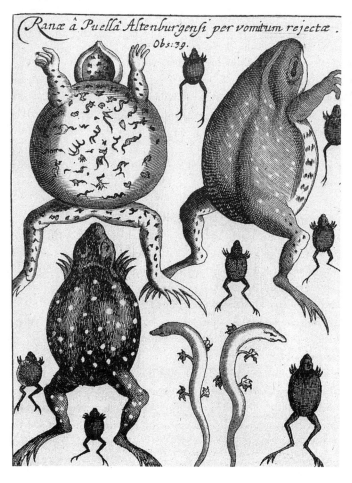

Fig. 4. Frogs and toads vomited by the Altenburg woman. From Thomas Bartholin's periodical *Acta Medica et Philosophica Hafnensis* of 1673. From the author's collection.

son. Every cough from the naughty boy brought his clerical guardians into a state of great excitement, and speculation was rife what monster would next make its appearance. Once, the clergymen thought they saw a large snake thrust its head out through the boy's mouth, but when the muscular Christians rushed forth, eager to "pull the Devil out and thrash the life out of him," it hastily withdrew to its lair among the boy's intestines.

In the meantime, a physician had dissected one of the vomited "frogs from Hell." It had several half-digested insects in its stomach, evidence that the frog had been alive outside the boy's body shortly before it was "vomited." The boy's attendants, who were in a state of excitement verging on hysteria, made no such deduction. They merely concluded that the frogs were supernatural and did not obey the ordinary laws of physiology. The clergymen experimented by pouring horses' urine, a time-honored cure for animals in the stomach, on the remaining live frogs. When the animals promptly died, it was unanimously decided to try this strong "medicine" to expel the devils that had turned the boy's intestines into a zoo-cum-hardware store. Poor Theodorus was made to swallow several large bottles of urine, and under the influence of solemn prayers and incantations of exorcism, this drug worked miracles. The boy stopped vomiting animals and objects. The clergymen exulted over the successful exorcism, and medical works of the time reported it. What seems more likely is that the threat of a second generous helping of horses' urine deterred the mischievous Master Theodorus from performing any further pranks.

By this time, several different theories had been formulated to explain how the animals reached the human stomach. Whereas most people believed that they were accidentally swallowed as spawn in polluted water, as in the Geisslerin case, others supported the age-old theory that snakes and lizards could crawl down the throat of sleeping people. Another hypothesis was that witches and sorcerers could torment their victims by putting reptiles inside them, as in the Döderlein case. To a historian of science, the most interesting of these theories is the famous doctrine of spontaneous generation, which flourished from the time of Aristotle well into the eighteenth century. Many odd notions existed about the generation of different animals: wasps were engendered from rotting horses' cadavers, geese grew on trees, and frogs, fish, and lemmings could rain down from thick clouds. It was generally accepted that parasitic worms could form from the corrupted humors of the intestinal tract: for some scientists, it was an easy step to propose that various other animals had the same ability.

Fig. 5. Young Theodorus Döderlein vomits a veritable torrent of newts, frogs, and other animals, as depicted in Georg Abraham Mercklin's thesis *De incantamentis* (1715). From the author's collection.

These imaginative theorists carried their notions to the most absurd lengths. Several bizarre case reports about shrews, moles, and cats in the human stomach were published in the notorious seventeenth-century German "monster magazine" *Miscellania Curiosa*. In 1721, the Russian court physician Dr. Klauningen described the case of a man who had vomited a living dog! When Herr Sabatius, a nobleman who was related to Field Marshal Scheremetjev, passed through the village of Sluck after the battle of Poltava, he felt unwell. The worm medicine given him by the village apothecary had no effect. Instead, he felt movements in his stomach, and later vomited a small hairless dog that, it was noted, looked rather like a pug. After its death, it was preserved in alcohol as a curiosity. Czar Peter the Great took it with him some years later after a visit to Sluck. More sensible colleagues did not believe the Russian doctor's tall tale about the spontaneously generated stomach pug: they claimed that the "dog" had instead been a large coagulum, and Dr. Klauningen was unable to refute them because of the czar's unwelcome interest in his specimen.

Linnaeus and a Cure for Frogs in the Stomach

During his famous journey to Lapland in 1732, Carl Linnaeus was consulted by the wife of the parish organist in the little town of Lycksele. She complained of having three live frogs in her stomach, which could clearly be heard croaking, particularly in the springtime. Linnaeus advised her to use liquid tar as a medicine to expel the animals, but she preferred large doses of strong aquavit, which at least kept the animals happy and quiet. In his medical conclusions to the *Iter lapponicum*, Linnaeus asserted that frog spawn, left floating in the brooks and streams by the adult frogs, was a serious threat to the health of the Laplanders. When they drank this polluted water, the spawn adhered to the membranes of the stomach, which "formed the nest, or rather the pond, for these dreadful animals, which tear and torture the poor patient."

In the late eighteenth century, the majority of the leading biologists, including Linnaeus, Georges Louis Buffon, and Johann Friedrich Blumenbach, supported the theory that snakes and frogs were able to live as parasites in the human gastrointestinal tract. In 1780 Sir Joseph Banks received a letter from the Reverend Samuel Glasse, describing how a certain Thomas Walker, after being given an emetic, had vomited a 2½-inch live toad, which crawled on the floor. In the early 1800s, more cases were published than ever before.

One worthy example, taken from the *Edinburgh Medical and Surgical Journal* of 1813, details the case of a maidservant employed by the parson of Dunfermline. According to Dr. John Spence, physician, the woman, "about 21 years [old], stout and firm in flesh," had suffered from vomiting, stomach pain, and constipation, and Spence had given her a strong dose of calomel. This medicine had the desired effect, but she was greatly frightened by passing a small lizard as well, which leaped out of the chamberpot and darted in under the drawers, from whence she could see its staring clear eyes. Undaunted, she pulled the animal out with a poker and flung it on the fire, where it expired with a shrill squeak.

Not until the 1830s was any frontal assault made on this age-old medical doctrine, and then not by any consultant or professor but by a mere German general practitioner. In 1834 a certain Dr. Sander was consulted by the thirty-year-old Frau Henriette Pfennig, whom he described as "eine kleine, sehr dicke und blasse Person, brünett, unregelmässig menstruiert, sehr ungebildet" (a short, very stout, and pale individual with dark hair; menstruates irregularly; quite uneducated). She had vomited two live frogs after being troubled by vomiting, stomach cramps, anxiety, and strong quacking noises from the midriff for more than a year. She said she had had an epileptic seizure in the marsh and, when she fell down, had imbibed polluted water containing frog spawn. During the following weeks, she vomited nine more frogs before an admiring crowd of spectators, several of whom had traveled for miles to see this celebrity and hear the frogs croaking within her. Whereas the specialists consulted were convinced that many amphibians still lodged in her stomach, Dr. Sander was skeptical. He dissected some of the frogs that Frau Pfennig had vomited and was greatly astonished to find several half-digested insects and an almost intact beetle in their stomachs. He also experimented on vomited frogs that were still living, finding that they were unable to live more than a very short time in the heat they would encounter within the human body. With these highly suspicious circumstances in mind, Sander went to the city magistrate, who bullied and threatened the poor woman severely. Weeping profusely, she admitted the hoax: after her neighbors had doubted that she really had frogs in her stomach, she had begun to carry these animals in the inside pockets of her skirt. In order to gain sympathy and public attention, she pretended to cough them up with enough skill to trick everybody present, including the medical men.

The clever Dr. Sander concluded his case report on the unfortunate Frau Pfennig by stating his conviction that all the other cases of snakes, lizards, toads, frogs, and fish in the human stomach could be explained as easily as this one. The defenders of the bosom serpent, however, rallied around a new case report, said to be the best authenticated ever. The Russian court physician Martin Wilhelm Mandt, who was very highly respected among his colleagues and enjoyed the full confidence of the czar and his family, had been consulted by a peasant who was certain that a snake had slithered through his mouth when he was sleeping out in the open. He awoke with a jerk and felt something cold moving around in his stomach. After taking a couple of stiff vodkas to calm himself and the snake, the peasant jumped onto his horse and galloped off to seek medical assistance. The noble ladies and gentlemen who were strolling in the park at Grand Duke Mikail Pavlovich's summer palace outside Oranienbaum were probably astonished to see the disheveled peasant, who was in a state bordering on hysteria, spurring his mount through the hedges and flower beds, and calling out loudly that only the foremost physician in the land could help him, since he had a snake in his stomach. Dr. Mandt managed to calm him down, and after he had been put to bed, performed a thorough examination of his abdomen. Mandt could feel movements in the epigastric area and hear a gargling sound with the stethoscope. The peasant was first given an emetic, but he vomited only slime. After four days, Mandt administered a strong purgative consisting of calomel, jalapa root, and castor oil. The man soon felt better, and the movements in his stomach ceased. He went home to recuperate, but returned triumphantly two days later, carrying a chamberpot containing the body of a twelve-inch adder (*Vipera berus*).

Dr. Mandt's case was widely published across Europe in translation, and even went the rounds of the newspapers. Since it was illustrated with a picture of the snake, his article did much to reestablish the old fallacy. But even this famous case crumbles before a critical reexamination. Anyone who has ever held a laryngoscope realizes how absurd it is to think that a serpent of this considerable size, as thick as a strong man's finger, could crawl through

Fig. 6. The stomach snake described by Dr. Mandt in 1839. From the author's collection.

the throat and gullet of a sleeping man without waking him. Nor it is possible for a snake to live four days in the human stomach. There were no witnesses to either the snake's dramatic entrance into the Russian peasant's body or its unceremonious expulsion from it. Dr. Mandt, noting that the reptile's backbone was broken at several sites, believed that the breaks had been caused by the powerful contractions of the bowel. What is more likely is that they occurred while the snake was being purposely killed. Furthermore, the snake's body would certainly have been more extensively digested if it had passed through the entire gastrointestinal system of a man. Certainly, Dr. Mandt's description portrays a man genuinely terrified, at least initially. Perhaps the generous helpings of various unpleasant medicines cured his delusion, but he felt the need to justify his fear by deceiving the doctor with the body of an adder.

Professor Berthold's Investigation

Many other German cases of living animals in the stomach were published during the 1830s and 1840s. Some pathologists agreed with Sander; others were impressed by Mandt's case. Professor Arnold Adolph Berthold, of Göttingen, was one of Germany's foremost physiologists. His experiments with transplantation of testicular gland tissue, which were published in 1849, made him one of the earliest pioneers of endocrinology. The next year, Berthold published a monograph that was to solve the burning question of the existence of living amphibians as parasites in the human stomach; it attracted much more popular attention than his endocrinological experiments and caused him a good deal more trouble. Berthold was astonished at the vast literature on this subject, which had accumulated for more than three hundred years. Nearly every German pathological museum of repute contained some snake, frog, or newt that had allegedly been vomited by some patient after living for years within the human body. Berthold obtained permission to dissect several of these specimens, and all had partially digested insects in their stomachs, strongly indicating that they had been deliberately swallowed shortly before being vomited. These findings inspired Berthold to assess by means of experiment whether frogs and lizards could live within the clime of the human body. He proved to his satisfaction that none of these animals could survive in 29°Centigrade water and that frog spawn putrefied at this temperature. Arnold Adolph Berthold did not hesitate to declare that the spawn of frogs and toads could not develop within the human body and that all reports in the older literature of newts and other amphibians living as parasites in the human body must be false.

Independently of Berthold, the American physiologist J. C. Dalton performed a series of similar experiments fifteen years later after some tall stories about slugs and lizards living as parasites in the human stomach had appeared in the popular press. Dalton demonstrated that slugs could not live in such an environment and that they were digested by the gastric juices of a dog. To clench the argument, he fed a dog with live slugs and newts, not allowing it to chew them; when the dog was killed an hour later, the animals were all dead and partially digested.

In 1854 Jonathan Smith, the gunner's mate of HMS *Hastings*, was suddenly taken ill while at Portsmouth. Two hours before his death, a nine-inch snake was seen to leave his body by way of the mouth. The provincial newspapers marveled at this strange occurrence, and Smith's shipmates made nautical jokes, likening the snake to a rat leaving the sinking ship. Several medical men and naturalists took exception to this vulgar error among the credulous. The ship's surgeon and several other witnesses asserted that it was a long roundworm, and not a snake, that had been expelled from the unfortunate sailor's body. A correspondent to the *Notes and Queries* deplored the British public's addiction to these idle tales and concluded that "it is a very common delusion among people affected by hypochondriasis, that they have swallowed reptiles in drinking ditch or pond water."

This final European case of a frog in the stomach occurred in 1882, when a lady patient of the Austrian practitioner Dr. Weiss claimed to have vomited a live frog after suffering from abdominal discomfort for some time. The naive doctor was quite convinced by her story, but when he published the case, his professional brethren protested that they had many times observed hypochondriacs complaining about snakes or frogs in the stomach "mit geradezu unerschütterliche Hartnäckigkeit" (with almost unshakeable obstinacy). The outcome was that the editor of the *Wiener Medizinische Blätter* had to apologize to the readers for lending credence to such an absurd old fallacy. Sigmund Freud was practicing in Vienna at this time, and it is likely that he was a reader of this periodical, but unfortunately he did not express his opinion on the psychological mechanisms involved in these delusions.

Although the bosom serpent had become a largely disregarded phenomenon in European medicine, it found new friends in America. In 1881 J. W. Anderson of Granbury, Texas, vomited "a snake twenty-one and a quarter inches long, of the prairie coachwhip variety," which had bothered him for six months; he threw it on the floor and his wife killed it. Although Dr. J. C. McCoy vouched for its authenticity in the *Texas Medical and Surgical Recorder*, adding that it was to be exhibited at the next State Association, it remains unproven that this bosom serpent was "the real McCoy." In another bizarre report, Dr. T. B. Fisher of Marion, Ohio, described the case of a lady who had felt something moving in her stomach for four months. She was ridiculed by her friends as a hysteric, but she silenced them by vomiting a nearly fully grown mouse, which Dr. Fisher kept in a glass jar in his office as a pet. As late as 1891, R. H. McCharles, M.D., of Cypress River, Manitoba, described "a patient whose ailment is of a most unique character," a farmer who had brought up a living frog after eighteen months of abdominal discomfort.

Eight years later, Professor Alfred Stengel of the University of Pennsylvania was much more cautious in describing two patients he had treated for alleged snakes in the stomach. He had tried to deceive one of them by sedating him while passing a ventricular tube; after the stomach had been washed out, a live snake was produced! The man was overjoyed, but some days later he complained that the snake must have had young while within him; he could feel them writhing in the intestines. He was then discharged from the hospital as incurable. After his death two years later, an autopsy revealed that he had suffered from what today would have been called chronic hypertrophic gastritis, a kind of inflammation of the stomach wall. Another individual, a thirty-four-year-old farmer, was admitted to Stengel's wards complaining of a snake in the stomach. His principal lament was that "he will never allow me to drink whisky; he hates that worse than anything else." The man had for some time been able to eat only bread; he was emaciated, nervous, and feeble of intellect. He was treated with a ventricular lavage, followed by free access to tonics and nourishing food. After a month, he had gained fifteen pounds in weight, and he was convinced that the sensation of a snake in the stomach had been a delusion. Stengel pointed out that many other presumed cases of living animals in the stomach amounted to misinterpretation by feebleminded people of the signs of gastritis or some other stomach or intestinal disease.

Bosom Serpents in Folklore

In large parts of Europe, stories of snakes, frogs, and lizards living in the stomach were passed on for generations among the village gossips. A large arsenal of popular cures against these internal parasites existed, and they were frequently put to the test. When, in 1864, a Laplandish peasant was sleeping on the floor in a crowded guesthouse, he felt something cold slithering down his throat. He awoke with a jerk, feeling sure that there had been a snake hidden in a basket of damp firewood brought in during the night, and that it was now inside him. He placed his confidence not in doctors but in the village quacks, who, in their turn, felt confident that they would be able to cure him. First, they tried a conventional method, giving him a laxative of Epsom salts and aloe, which only irritated and enraged the snake. They next resorted to the famous old nostrum of horses' urine, but without success. Then both the snake and the stomach lining of its host withstood the corrosive powers of a strong solution of nitric acid. Finally, the man swallowed the strongest serpentofugic drug available — a mixture of sour small beer and tobacco oil from old uncleaned pipes — but the bosom serpent put up with this very unpleasant meal and remained in its lair.

The next step was physical therapy. Two strong men, one on each side of him, applied their hands to the large snake bump on his belly; after a violent massage, lasting nine hours, the snake was knocked out cold for more than a day before recovering its senses. The most bizarre household remedy still remained, however; the quacks made a strong hook from steel wire, and baited it with a lump of bread. After half an hour's angling, the snake bit the hook! There was great rejoicing among the interested spectators, and the patient groaned out that he felt the bosom serpent being hauled through his intestines and up the gullet. When it had almost reached his mouth, he felt it falling down into his stomach again! The empty hook was rebaited and swallowed, but the serpent was less eager to bite it. The intrepid snake fishermen did not give up even when the hook got stuck in the stomach wall; they managed to free it somehow, and they continued their angling until the patient fell unconscious from the strain and exhaustion. Only when all these cures had failed did the peasant consult the district general practitioner, who listened to his long tale of woe with increasing horror. No rational arguments could shake the man's conviction that he really had a snake in his stomach, and no doctor or quack could help him.

In German and Scandinavian folklore, the snake, as a close ally of the Evil One, was able to hypnotize careless forest ramblers and thus achieve a comfortable boardinghouse existence within them. Or a snake might invade a sleeping child in its cot. Swedish haymakers always took a pinch of snuff before falling asleep in the fields, hoping that the snakes and slowworms would not desire to negotiate such a foul-tasting obstacle. In Scandinavian mythology, stomach frogs were also much feared, since they were believed to grow to an enormous size. In a tale from Norway, the frog became as large as a German shepherd dog, and the wretched patient could feel its claws closing in around her heart before she died.

Early in 1916 a remarkable story appeared in several English newspapers. A woman had swallowed frog spawn, which had developed into a large frog that lived and grew inside her. She was taken to Stroud Hospital (in Gloucestershire), but the doctors were unable to operate, since the animal moved about too fast. The woman was in such agony that the baffled medical men wrote to King George V for permission to kill her with poison, but his majesty refused this plea for euthanasia. Instead, as a last resort, the doctors put a piece of cheese on her tongue, and the frog smelled it and jumped up her gullet to eat it. Tragically, the obese amphibian was too big to get through, weighing half a pound, and the wretched woman choked to death.

This tale somewhat resembles the old Yorkshire belief in a "water wolf," which could be drunk in impure well water. The "wolf" resembled a lizard or toad, and could grow to an enormous size within the body. The end was near when it could be felt to use the heart as a headrest while going to sleep in the evening. According to a newspaper story in the *Keighley Herald* of 1909, a certain Miss Judson of Haworth (the home of the Brontë sisters) had fallen foul of the dreaded water wolf. She could feel it moving and kicking inside her. She then prepared a "tasty bit" consisting of butter, onions, and salt, mixing these ingredients together in a pan. The vegetarian water wolf apparently shared her preference for this delicacy, since it leapt up her throat and out through her open mouth. Miss Judson had the presence of mind to shut her gaping jaws, since she had heard of another woman who had got rid of a water wolf in the same manner, only to have it jump back in again. The water

Fig. 7. An illustration from Archbishop Olaus Magnus's *Historia de gentibus septentrionalibus*, showing the reason why small children's cots were strung from the trees in Scandinavia in the early 1500s. The snakes gaze wistfully from the ground, but are unable, because of this wise precaution, to achieve a comfortable life inside the child. From the author's collection.

wolf landed on the floor with a heavy thud and there encountered Miss Judson's fierce cat. In apparent danger of becoming a "tasty bit" itself, the water wolf committed suicide by leaping into the nearby fire, giving off a stench of bad eggs as it perished.

In 1921 another ludicrous story appeared in the Norwegian newspapers. An adder had crawled through the mouth of a young country girl, and although the nation's foremost physicians had been consulted at the Rikshospitalet in Oslo, they had pronounced her incurable, since the reptile had had young inside her. King Haakon had been petitioned for permission to kill her as an act of mercy, but the king refused, saying that he had so few subjects that he wanted to keep every one of them alive. The story spread like wildfire all over Norway, and the entire nation pitied the poor girl with the family of adders within her. A week later, the director of the Rikshospitalet denied the whole story. It was completely untrue. More than likely it had grown out of an incident in which a long ascarid was found during an appendectomy at the hospital, and the story had grown more and more dramatic on its way to the newspaper headlines. In a curious sequel, not fewer than eighteen letters were sent to the hospital by credulous country people, suggesting various household cures for the unhappy girl. An old woman recommended that the girl was to breathe the aroma of newly baked bread, for this was an infallible means of coaxing the hungry snakes up from their lair. Her own mother had suffered from snakes in the stomach without knowing it, and one day, after she had fallen asleep while baking, her husband was aghast to see seven snakes crawling about on the loaves of bread in front of his sleeping wife. The rough Norwegian killed the snakes and placed them in a row before his wife, waking her with the words: "Here are the children you have just given birth to!" The poor woman had a fit and died!

The most famous literary snake in the stomach appears in Nathaniel Hawthorne's short story *Egotism; or, The Bosom Serpent,* which was originally published in 1843. The hero of this story, Roderick Elliston, carries a snake in his stomach, which is a symbol for his own diseased jealousy and self-centeredness. Like a man deranged, he roams the pavements of his hometown, clutching his breast and muttering, "It gnaws me! It gnaws me!" He freely accosts passing strangers, accusing them of having their own bosom serpents as well, telling a quarrelsome clergyman he had swallowed one in a cup of sacramental wine, and a miser that he had engendered a copperhead snake within him by defiling his fingers with base metal. He asks a stout and ambitious statesman about the welfare of his boa constrictor. After Roderick is reunited with his wife, he is able to rid himself of the sensation of a reptile gnawing his heart.

Hawthorne probably got the idea for his story from the contemporary American newspapers, which had no shortage of articles about people with animals in the stomach during this heyday of the bosom serpent legend. In 1836 Hawthorne was already making some notes on this topic for future use. Another famous American writer who was interested in the legend of the bosom serpent was Henry David Thoreau, who wrote in his journal about the fear of swallowing a snake's ova in brook water. He recommended that the bosom serpent was to be coaxed up from its lair at the sound of running water, and then caught with a powerful grip round the head and pulled out, even if it might feel as if its tail was coiled around the patient's vitals.

In 1991 when I was visiting a psychiatric ward to examine a patient with heart trouble, one of the medical students there, who knew of my interest in these matters, asked if I wanted to see a woman with a snake in her stomach. She was a very stout old woman, who had been admitted some days earlier with severe delusions and hallucinosis. She said that a wicked clergyman who ran a refuge for alcoholics had deliberately fed the snake to her as spawn more than a year ago, in a bowl of soup. She had become used to her bosom serpent by now, and the worst thing was that she had recently got a small computer in the stomach as well; she could hear it buzzing and beeping, as if the snake were diverting itself by playing computer games. I asked how on earth this machine had got into her, and she answered, "The bloody parson fed it to me in the same way!" This comment conjured up a weird picture of a swallowed microchip in the soup, dividing and subdividing until it had achieved megabyte capacity.

The Bosom Serpent Unveiled

The belief in snakes, lizards, and amphibians being able to live for prolonged periods of time in the human stomach can be traced back to the dawn of history, and in ancient legends from different parts of the world, tales of these bizarre parasites are recounted with a singular fascination. I discovered not less than sixty-eight case reports of live reptiles or amphibians in the human gastrointestinal tract on record in the medical literature, evenly distributed through the seventeenth, eighteenth, and nineteenth centuries. More than forty of them are from the German states, where many medical men remained firmly convinced of the bosom serpent well into the 1840s.

Seven cases were from France, four from Scandinavia, and only two from Great Britain, where the old fallacy never received much support among the learned. The five American cases are all from the later part of the nineteenth century. Of the sixty-five stomach parasites reported, there were sixteen snakes, eighteen lizards, twelve toads, fourteen frogs, two mice, two dogs, a cat, a shrew, a mole, and a hen. Thus, the snake in the stomach was less commonly observed than the lizards and amphibians, but in folklore, it was far more revered.

Although some skillful early clinicians, such as Alexander of Tralles and Ambroise Paré, were able to see through the convincing stories and dramatic symptoms of their patients, the man who did most to disprove the ancient fallacy of the snake in the stomach was Arnold Adolph Berthold. Using the pioneer work of John Hunter and Lazzaro Spallanzani on the temperature regulation of cold-blooded animals, he was the first to demonstrate conclusively that amphibians and their spawn could not long remain alive inside the human body. His investigation may well have been inspired by the clever practitioner Dr. Sander's revelations. Like Sander, he dissected the vomited animals and found proof they had recently lived outside the human body. Since Berthold had probably also read and appreciated John Hunter's pioneer work on the nature of the gastric juices, it is odd that he did not consider the effects of digestion on the alleged parasites. This line of inquiry was left to the American physiologist J. C. Dalton fifteen years later. Berthold's countrymen were quick to appreciate his findings. From the year 1850, when his monograph was published, to the present day, not a single case was reported in the German periodicals, which had earlier presented many a sad example of physicians' gullibility and sensationalism. In northern European folklore, however, tales of live animals in the stomach were retailed well into the twentieth century. When my own mother was a little girl in the 1920s, she was solemnly warned by a well-meaning aunt not to drink water in stagnant pools and ditches, for fear of accidentally imbibing frog spawn, with the most dire consequences.

The bosom serpent is still with us, incorporated into modern newspaper myths. In 1979 the *Daily Star* gave credence to the story of a fifteen-year-old Turkish girl with three long water snakes in her stomach, which had lived there for five years after being accidentally imbibed as eggs. Another London paper, the *Daily Telegraph*, went one worse in 1982. A twenty-five-year-old Syrian woman was taken to the hospital complaining of severe stomach pains. It was ascertained — perhaps after the surgeon had met the gaze of its reptilian eyes through the fiber optics of his endoscope — that she had a six-foot snake lodged in her intestines. According to a Syrian newspaper, it made a sound like a chicken when hungry, loud enough to be heard by people standing near her. The Syrian surgeons failed to remove the serpent, and the patient was sent to Spain, where a second operation also failed, even though the anaesthetic was used on the snake as well as the patient.

Another ludicrous canard went the rounds of the newspapers in 1987. An eleven-year-old girl in Baku, Azerbaijan, had a twenty-six-inch "semi-poisonous Caucasian cat snake" in her stomach, which had slithered down her throat while she slept. The clever doctors managed to flush it out with the stomach pump. The *Scotsman* added a further *frisson:* she was still in the hospital, being treated for the enraged reptile's bites in the stomach wall. Inquiries to the hospital in question, made in 1990, revealed that no one there recalled this remarkable case; the hospital clerk rather suspected that a hoaxing letter to *Pravda* might have started this silly story on its way around the globe.

Some medical historians have advanced the hypothesis that belief in the stomach snake originated when simpleminded people mistook large round-worms (*Ascaris lumbricoides*) for snakes. These parasites are not exactly snake-like, but they can grow to be three or even four inches long. That the "snakes" vomited were in fact ascarids is apparent in some early cases, but not in any later ones, and of course, this hypothesis is completely inadequate to explain other animals named as parasites. During the eighteenth and nineteenth centuries, the fear of having live snakes or frogs within them touched people to the core. It seems likely that individuals with some organic or psychosomatic gastrointestinal disease were sometimes seized by the delusion that they had one or more living animals in their stomachs. The protuberance in the hypogastric region which many patients interpreted as "one of the snake's coils" might well have been a flatus-filled section of the large bowel. This "internal delusion of infestation" might sometimes progress so far that the patients pretended to vomit living or dead amphibians, presenting them as evidence that their stomachs were still full of such animals. Some of the frog vomiters were obvious frauds or hysterics; others were alcoholics who had chanced upon this clever way of convincing people that they needed strong liquor as a medicine to calm down their unruly passengers within. In nineteenth-century Germany and Scandinavia, there were even some quacks or snake-doctors who specialized in ridding people of animals in their stomachs. During treatment, the hypochondriac sat with jaws wide open, while

50   the quack searched for snakes and frogs; today, people sit in the same position for the measurement of electric currents, auras, and magnetic fields. Just like standard medical science, the realm of imaginary medicine has advanced technologically in the twentieth century. The uncomplicated old medical delusions have gradually died away, and the place of the bosom serpent has been taken by complicated chemical and electromagnetic theories, odd psychological notions, and ancient oriental philosophy.

The Riddle of
the Lousy Disease

  ❧ *O shamefull Plague! O foul infirmitie!*
Which makes proud Kings, fouler than beggars
    be
(That wrapt in rags, and wrung with vermin
    sore
their itching backs sit shrugging evermore)
To swarm with Lice, that rubbing cannot rid,
Nor often shift of shirts, and sheets, and bed:
For, as in springs, stream stream follows fresh,
Swarm follows swarm, and their too fruitfull
    flesh
Breeds her own eaters, and (till Death arrest)
Makes of it selfe an execrable feast.

From Josiah Sylvester's translation of Guillaume du
Bartas's famous poem *La semaine* (1605)

O F ALL THE LEGENDARY AND FANTASTIC diseases of ancient times, phthiriasis, or the lousy disease, was the most intriguing and bizarre. In the corrupted humors of those who suffered from the disease, lice were believed to be spontaneously engendered, and tumors full of these insects, but devoid of pus, rose on the skin. When such a louse tumor burst or was incised, a stream of insects swarmed out. The flesh of the sufferer was slowly eaten away and transubstantiated into lice, and he perished miserably in this "most horrible of diseases." Phthiriasis was firmly believed to be a divine punishment for tyrants, desecrators, and enemies of religion. Belief in the curse of the lousy disease was well established in antiquity, and it lived on until the early 1800s. In the medical literature, phthiriasis was accepted even longer, and many cases were reported far into the nineteenth century. Not until the 1860s was it finally abolished from the textbooks after a fierce debate among some of the leading dermatologists of the time.

Phthiriasis in Antiquity

The annals of phthiriasis stretch far into time. One of the earliest descriptions of the disease was given by Aristotle in his *History of Animals*. Lice, he believed, were produced from the flesh of the human body and gathered in small eruptions on the skin. When these eruptions were opened, a mass of lice emerged, but no purulent fluid. About a hundred years later, in 240 B.C., the geographer and historian Antigonos Carystius described a similar disease; lice were formed in the flesh, and when the insect-filled nodules under the skin were opened, they swarmed out. He seems to have been the first to call the disease *morbus pedicularis*, a name it was to keep for more than two thousand years.

Another curious account of phthiriasis is given by the historian Diodorus Siculus about 50 B.C. A North African tribe of locust eaters very often died of phthiriasis, breeding in their bodies a peculiar type of savage winged lice. Itching skin eruptions first appeared on the breast and stomach and soon spread all over the body. When such an eruption was scratched, a multitude of these insects burst forth. The tissues of the sufferer were slowly eaten away, and the insects emerged from many small holes in the skin. Diodorus speculated whether the people's strange diet or the hot climate might be the cause of this endemic disease.

Caelius Aurelianus, in his work on acute and chronic diseases, also recognized the difference between ordinary lice and the "wild" lice of phthiriasis, which entered the tissues of the body. Hippocrates never mentioned the lousy disease, but it is discussed several times in the works of Galen, who believed that lice could be formed deep within the skin and create rather large boils. Like Aristotle, he attributed the disease to an excess of warm moisture in the body. The elder Pliny also discussed phthiriasis in his *Natural History*. Insects, he asserted, were formed in the blood of the patient and ate up his or her flesh. As a treatment, he recommended rubbing the whole body with juice of the Taminian grape or with hellebore juice and oil.

In his *History of Animals*, Aristotle mentioned two famous men who had died of phthiriasis, the Greek poet Alkman and the Syrian philosopher Pherecydes, the teacher of Pythagoras, who lived in the sixth century B.C. According to Diogenes Laertius, there were several stories about the death of Pherecydes. In one version he was buried on the battlefield of the Ephesians and the Magensians; in another, he hurled himself from Mount Corycus; and in a third, he simply died a natural death. The most popular version was that he died of the lousy disease; when Pythagoras inquired how he was doing, the lousy philosopher thrust his finger, swarming with vermin, through the doorway, and exclaimed, "My skin tells its own tale." This remark became a byword for "getting worse." There is also a spurious letter from Pherecydes to Thales, in which he bewails his fate with the words: "I am infested with vermin and subject to a violent fever with shivering fits." In his *Liber medicinalis*, the poet Quintus Serenus Sammonicus mentions the wretched philosopher's lousy death with the words:

> *Noxia corporibus quaedem de corpore nostro*
> *progenuit natura uolens abrumpere somnos,*
> *sensibus et monitis vigiles intendere curas.*
> *Sed quis non paueat Pherecydis fata tragoedi,*
> *qui nimio sudore fluens animalia taetra*
> *exsudit, turpi miserum quae morte tulerunt.*

This passage could be translated in the following way:

> *Nature sometimes creates certain noxious vermin,*
> *dangerous to the body, out of our own bodies,*
> *in order to break up sleep and, in inciting our senses,*
> *sharpen our attention.*
> *But who does not fear the fate of the tragedian Pherecydes,*
> *whose copious sweat contained loathsome little animals;*
> *they brought him to a heinous death.*

Pherecydes was by no means the only philosopher to perish by phthiriasis. According to Diogenes Laertius, lurid proverbs about "Plato's lice" soon began to circulate after the great philosopher's death, and Plato's nephew Speusippos was also said by some to be a victim. It is most likely that these accounts were invented by enemies of the Academy. Even Socrates and Democritos were said to have been "eaten by vermin," certainly with even less foundation.

Although Aristotle and Galen do not appear to have considered the lousy disease a punishment for transgressors, it is obvious from these highly dubious accounts of the lousy Greek philosophers of antiquity that dying in phthiriasis implied a moral stain on the deceased. This taint is even more apparent in the accounts of the death from phthiriasis of Herod the Great and Herod Agrippa in Flavius Josephus's *Antiquitatum judaicarum* and the Acts of the Apostles. After Herod Agrippa had been hailed as a god, "an angel of the Lord smote him because he did not give God the glory, and he was eaten by worms and died." Another victim of the curse of the lousy disease was the Syrian king Antiochus IV Epiphanes. According to the second book of Maccabees, the body of King Antiochus swarmed with vermin while he was alive, and his flesh rotted away. His army was revolted by the stench of his decay. The king promised to enrich Jerusalem and convert to Judaism, but to no avail, and he died miserably of this disgusting malady. In his *Canterbury Tales*, Geoffrey Chaucer describes his death with the following lines:

> *The wreche of god him smoot so cruelly*
> *That thurgh his body wikked wormes crepte;*
> *And therwithal he stank so horribly. . . .*
> *No man ne myght hym bere to ne fro,*
> *And in this stynk and this horrible peyne,*
> *He starf ful wrecchedly in a monteyne.*

The most famous of all ancient chronicles of phthiriasis appears in the *Life of Sulla*, in which Plutarch lists a fair number of victims of this disease. The earliest is Acastos, an enemy of the father of Achilles. This legendary figure was said to have lived as early as the eleventh century B.C., but some historians have doubted his existence altogether. Nevertheless, it is apparent that the tradition of phthiriasis existed long before Herodotus and Aristotle gave their accounts of the disease. According to Plutarch, the jurist Mucius Scaevola and Alexander the Great's treacherous henchman Callisthenes both died of phthiriasis, as did Eunus, the leader of a slave rebellion. The *historia morborum* of Sulla, the best known of all historical victims of the lousy disease, was described in hideous detail. The tyrant's corrupted flesh became one mass of lice, and although many men were employed to remove and wipe away the vermin, they continued to multiply, and his clothes, bath, furniture, and food were full of them. He bathed frequently and every day washed and rubbed his body, but to no avail. So rapidly was his body transformed into lice that all attempts to cleanse him were frustrated. Pausanias also gloated over this suitable death for the Roman tyrant, and Pliny made an ironic comment on Sulla's unsuitable epithet *felix*: were not his victims more fortunate in dying than he, asked Pliny, "when his body ate itself away and bred its own torments?" The poet of phthiriasis, Quintus Serenus, wrote:

> *Sulla quoque infelix tali languore peresus*
> *corruit et foedo se uidit ab agmine uinci.*

This passage could be translated as follows:

> *Sulla the Unhappy also perished in this painful disease;*
> *he was defeated and vanquished by an infamous host of enemies.*

The Curse of the Lousy Disease

In Plutarch's time most writers agreed that phthiriasis was a punishment from the gods against highly placed men who had offended them. For example, Quintus Pleminius, the legate of Scipio Africanus, who had plundered the temple of Proserpine, was struck down by this disease. The early Christians eagerly took over the myth of phthiriasis, often using it to denigrate fallen enemies. When one of the last great Roman persecutors of their faith, the emperor Galerius, died in 312 B.C., the Christian apologists Lactantius and Eusebius spread the rumor that he had perished of the lousy disease. The pious Lactantius described the emperor's grisly end with gusto: the tyrant rotted from within, generating vermin that ate the flesh from his bones; his legs and lower body were swollen and putrid, while his upper torso was withered and mummified. The emperor was tormented in this way for more than a year before he acknowledged God. The apologists also exulted in the horrible fate of the emperor Maximinus Daia: his eyes popped out of their sockets, he went mad, and his body was desiccated to little more than skin and bones; the most bloodthirsty of them added that he finally perished of the lousy disease.

Until about 1850, neither medical men nor historians doubted these astounding tales of the lousy great men of antiquity, but in later years several classical scholars have been skeptical. Closer study of the sources and the historical background certainly gives rise to doubts about this bizarre epidemic of phthiriasis among the tyrants of classical history. In the case of Antiochus IV, both the author of Second Maccabees and Flavius Josephus got their grisly details from a no longer extant source, the chronicles of Jason of Cyrene. His dark portrait of Antiochus was probably inspired by Herodotus's stories of the megalomaniac King Xerxes and the lousy Queen Pheretima of Cyrenaica. More realistic historians mentioned neither the fatal hubris of Antiochus, which called down the wrath of the gods, nor his lousy death. Plutarch got his gruesome details of Sulla's deathbed from the *Populares*, a contemporary chronicle of gossip and polemic, which was more remarkable for sensationalism than for reliability. The story suited his own theories of moral and divine retribution, and he preferred it to more authentic reports of the tyrant's last days. According to the chronicler Appius, Sulla was in good health and died suddenly of a stroke. Plutarch undermines his own credibility by maintaining that Sulla, while beset by this horrible agony, received his friends, took care of his correspondence, and even finished his memoirs two days before his death. The case of Herod the Great was built up in a similar way: his opponents depicted him as a rotting monster, swarming with vermin, ordering rabbis to be burned alive from his deathbed, and executing his son Antipater; the court historian Nicolaus of Damascus, who was in a position to know, mentioned nothing of these dramatic excesses.

It is obvious that the partisan historians and chroniclers of antiquity often had little foundation for their grisly and sadistic accounts of the death agonies of the tyrants and immoral desecrators. It seems to have been common in political and religious propaganda to spread the rumor of a fallen enemy's death in phthiriasis, thereby implying that higher powers had been against him and his cause. Not even Judas Iscariot escaped this desire to accord evildoers a suitable death: according to Matthew, he hanged himself, but the Acts of the Apostles says that "he fell headlong, and his body was shattered so that his entrails poured out." Bishop Papias of Hierapolis considered even this death too mild for the wretched Judas: he wrote that the traitor's body bloated to such enormous size that it could not even be brought through a gate built for wagons. His eyes were swollen shut, and his genitals were the most disgusting sight imaginable; from every part of the body oozed a stream of pus and vermin. So powerful was the stench, the pious bishop insisted, that no one to this day could pass the spot where he died without holding his nose.

The idea of disease as a punishment for sin is extremely old, and it can still sometimes be encountered among individuals with a disposition toward religious brooding; it reached its most extreme form in the curse of the lousy disease. Susan Sontag in *AIDS and Its Metaphors* says that this idea was not particularly widespread during antiquity and that she found it only in Sophocles' tale of the archer Philoctetes and his foul-smelling wound. This is quite wrong, however, and a closer study of the myth of phthiriasis might have been of considerable interest for the discussion in this work.

A Royal Malady

From the Middle Ages there are few reliable medical records of phthiriasis, but the legend of the curse of the lousy disease was very widespread and appears in many historical chronicles. Theologists, moralists, and historical compilers delighted in constructing long lists of the hapless victims of this disease; the most detailed being those of Theodor Zwinger, Ulysses Aldovandri, and Thomas Bartholin. The Saxon nobleman Radbertus, who had treacherously slain Bishop Praejectus of Clermont, and an uncle of Emperor Julian the Apostate, who had desecrated the high altar in Antioch by urinating on it, were two prominent victims of this royal malady, as were the excessively vain emperor Arnulphus of Franconia and the villainous Medicean antipope Clement VII. The curse of the lousy disease also smote two sinful bishops guilty of simony: Lambertus of Constanz and Fulcherus of Nymwegen both paid for their crimes in this way. Fulcherus was said to be so full of vermin that he had to be buried sewn into a deerskin sack. The Vandal king Huneric, who had exiled 444 Christian bishops, was another victim of phthiriasis; his many crimes inspired the Christian historians to excel in cruel accounts of his death. Isidorus in his chronicle of the Goths reported that his entrails poured out, and Gregory of Tours claimed that he tore himself to shreds with his own teeth; Zwinger got the story about his lousy death from the chronicler Sigebertus.

The English medieval chronicles contain several notable cases of phthiriasis. The nobleman Leostanus, who doubted that King Edmund's hair and nails had grown after death and demanded that the corpse be shown to him, went mad at the sight of it and subsequently died in phthiriasis. During his life Ælfhere, ealderman of the Mercians, was an enemy of the monks, and after his death in 983 various Latin writers blackened his character. William of Malmesbury accused him of having been involved in the murder of King Edward the Martyr, whose body he buried, and also of having been consumed by vermin as a divine punishment. According to the Annals of the

58   Four Masters, Diarmaid MacMurchada, king of Leinster, also died of the lousy disease, his body becoming putrid while living "through the miracle of God and the Saints of Ireland whose churches he had profaned and burnt." King Fairchair (or Ferchardus) II of Scottish Dalriada also died of this royal malady, according to George Buchanan, who painted a dark portrait of this monarch.

Another royal case of phthiriasis is described in an old Danish legend of obscure origins. After the death of King Halvdan of Denmark, his kingdom was conquered by the Swedish King Adils, who appointed his favorite dog, Racke, Danish viceroy in order to mock his defeated enemies. One day, however, the dog king jumped down from his throne in order to separate two hounds who were fighting in a corner of his great hall, whereupon his canine majesty was promptly bitten to death by his sturdy opponents. None of the Danes was willing to bring the news of the dog king's death to King Adils, who had previously promised that the messenger in question would lose his head. After some time, a giant named Lae managed to persuade his cowherd Snyo to bring the mournful news, and by the use of cryptic answers to the king's questions, he managed to convey the message without incriminating himself. King Adils was so impressed by Snyo's cleverness that he appointed him to succeed the hapless dog as viceroy of Denmark. During his short reign, King Snyo was cruel, vindictive, and generally hated by his people. One day, he sent a man named Rödh to his old master Lae, ostensibly to ask the giant how Snyo would die when his time came. His ulterior motive was the hope that the fierce giant would tear this uninvited visitor to pieces. Rödh, however, managed to escape alive. He returned to the royal court at Jutland and announced the giant's sinister prophesy: King Snyo would die "from the bite of lice." At once, innumerable lice were seen to crawl forth on Snyo's body, and he died after a few days of torment, forever after to be known as the louse king. Historians agree that King Adils reigned during the sixth century B.C., but Snyo probably never existed. The tale of the dog king, without the additional embellishment of the lousy viceroy, exists in many variations in Nordic mythology.

Phthiriasis Strikes King Pym

The first reliable cases of phthiriasis are from the sixteenth century. In 1556 the Portuguese physician Amatus Lusitanus described the death of the nobleman Tabora, who had many swellings all over his body, from which small insects streamed out incessantly; two of his Ethiopian slaves were employed in emptying small baskets of them into the sea. After some weeks, he was devoured by these "lice" engendered under his own skin. Three more patients were presented by Petrus Forestus; one of them, a young painter's apprentice, had a large, itching boil on his back. When it was opened, a huge number of insects streamed out, but there was no pus or fluid; the man was cured of his phthiriasis. This was the first case of phthiriasis with only one large insect-filled tumor. Forestus himself, who had seen several cases, wrote that death usually ensued when lice gathered in swellings all over the body, but that the disease could be cured by opening an insect tumor such as this one.

Another remarkable case was reported by Thomas Moufet in his *Insectorum theatrum*. An English noblewoman, Lady Penruddoc, developed hundreds of small insect-filled boils and perished in phthiriasis. Moufet noted that the insects resembled mites rather than lice; unfortunately, he left no drawing of them. The medical writers of the sixteenth and seventeenth centuries did not consider the idea of insects forming beneath the skin in any way abnormal, since this was the age of spontaneous generation, when it was firmly believed that many kinds of animals could be generated from putrid flesh.

During the seventeenth century, phthiriasis was regularly mentioned in medical and entomological textbooks and collections of case reports. The medical men of this time were quite aware that ordinary *phthiriasis vulgaris*, of which almost all of them had had personal experience, was very different from *phthiriasis rara et horrenda species*, the rare and horrible lousy disease. Although it was believed that lice were engendered from human sweat by spontaneous generation, some people were presumed to have especially diseased blood, permitting the insects to burrow into the skin and live there in great numbers. In 1678 the first of several doctoral theses on phthiriasis was written by Georg Franck von Franckenau of Heidelberg. He defined phthiriais as the dissolution of some part of the body, due to the formation of a copious amount of lice therein, accompanied by a persistent fever and other symptoms. Although he quoted Francesco Redi's famous work disproving the spontaneous generation of insects, Franck von Franckenau put more reliance in the authority of the old supporters of this ancient doctrine, such as Aristotle, Girolamo Cardano, and Athanasius Kircher, than on such novel research. The cause of the formation of lice in phthiriasis could be corrupted blood and flesh, but supernatural causes could not be ignored. He made a long list of historical cases and tried to penetrate the Almighty's reasons for striking them down with the "most loathsome of diseases."

The historical annals of the sixteenth century had added several new cases. One of them was a controversial French statesman, Chancellor Duprat. Another, King Philip IV of Spain, whose case is of particular interest, since several details in his *historia morborum* bear some resemblance to the descriptions of phthiriasis in the medical literature, without the gloating exaggerations so common elsewhere. The aging king, who was said to be already in extremis from severe dropsy and gout, developed several abscesses on the chest and knee, and when these were opened, insects streamed out instead of pus. Several impartial historians have accepted the old legend that King Philip died of phthiriasis, but for the reasons already stated, all cases involving great men must be treated with extreme caution.

Had Franck von Franckenau been better informed about modern British history, he could have added another remarkable case to his collection. After the death of the parliamentary leader John Pym in 1643, the royalists delighted in spreading the rumor that, as a punishment for his disrespect toward King Charles I, God had struck him with "that loathsome and ignominious disease, called by Physicians, *Morbus pedicularis*." In order to scotch these rumors, his political allies had the corpse autopsied. Although the autopsy clearly showed that Pym had died of gastrointestinal cancer and although the corpse of "King Pym" was publicly shown in Westminster Abbey, where many hundreds of people saw it, the rumors persisted that he had perished of the "foul disease of Herod." The Cavaliers even had a cartoon printed in which the politician was pictured full of vermin, with the caption "Les pouls ont mangé Maistre Pin!"

Not the least curious of the many cases of phthiriasis in the seventeenth and early eighteenth centuries was the one presented by Michael Valentin in 1730: a forty-year-old man had a number of small, itching swellings dispersed over his body. After medication proved futile, one of the swellings was cut open at the patient's request. The man almost fainted from fear when countless lice burst forth, but once the tubercles were all opened and emptied of insects, the man recovered completely from his phthiriasis.

An interesting Swedish case from the same time was described by Professor Johan Lindestolpe of the Nosocomium Hospital in Upsala. A sailor had been admitted to this hospital in a cachetic state, with wounds and insect-filled swellings all over his body. Through treatment with mercurial ointment, Lindestolpe managed to cure the man completely, and he returned to his ship "blessing the hospital and the advances of modern medicine." Carl Linnaeus probably studied this famous case, since in the original Swedish version of his *Lectures on the Animal Kingdom*, he wrote that it is the worst kind of physical impurity when the lice build nests for themselves between the skin and the flesh; in order to cure this phthiriasis, or "louse-fever," as he called it, Linnaeus also recommended mercurial ointment with an addition of sabadilla seeds.

Dr. Alt's Thesis

From 1730 to 1802, no new case of phthiriasis was published, and at the end of this period, several men of observation questioned the existence of the disease. The entomologists now knew a good deal about the anatomy and physiology of lice, and they doubted the capacity of these aerobic insects to live under the skin and lay eggs there. Although some medical men still advocated the theory of spontaneous generation, it had little support from the entomologists and scientists of this time. In Britain both Robert Willan, in his *Description and Treatment of Cutaneous Diseases*, and William Kirby and William Spence, in their *Introduction to Entomology*, denied the existence of subcutaneous lice. None of these writers doubted the existence of the disease, however, and Kirby and Spence were among the first to propose that it might be caused by some unknown species of mites. They quoted a case from William Heberden's *Commentaries on the History and Cure of Diseases*, which he had in his turn obtained from Sir Edward Wilmot, who had examined a man afflicted with phthiriasis. Small, itching tumors were dispersed over his skin. Remarkably enough, there was a very perceptible motion in them. When opened with a needle, they proved to contain insects resembling common lice, except that they were whiter.

In France not even a revolution could shake the firm adherence to the old doctrine of the *maladie pédiculaire*. Two new doctoral theses, by Doctors Reydellet and Tournadour, were written in the early 1800s, but without adding much that was new. They also reported some new cases of alleged phthiriasis, but from the descriptions, it is obvious that the patients suffered from mere lousiness, often with some additional secondary infection.

The Prussian military surgeon Professor Rust reported a remarkable case that fit the classical symptoms, however. When he visited Prince Sangusko of Wolhinia in 1808, the town surgeon, Dr. Müller, asked him to examine a thirteen-year-old Jewish boy with a large head tumor, which was neither inflamed nor fluctuant. When Rust saw the boy again eight days later, he seemed to be dying, and the tumor was enormous. Rust considered it prudent to cut into the growth to find out what its contents were. To the horror of all present, it was found to contain a mass of solidly packed insects,

62   but not a droplet of pus or moisture. After the insects had been scraped out, the boy's head was rubbed with Neapolitan ointment, and the cavity of the growth injected with mercury. After a while, he recovered completely.

In 1824 the German Dr. Henric Christian Alt proposed a new theory in his doctoral dissertation on phthiriasis. He believed that a previously unknown species of louse, *Pediculus tabescentium*, or the phthiriasis louse, caused this disease, and that it did not develop from nits like the other lice but was spontaneously generated. Alt's theories were accepted throughout Europe, for they were generally considered a better explanation of the many bizarre features of the disease than those previously essayed. One of Alt's supporters, a German general practitioner named Stegmann, developed his own bizarre views. He declared that pederasts and other morally inferior individuals had an inherent tendency to phthiriasis; because of their immoral prac-

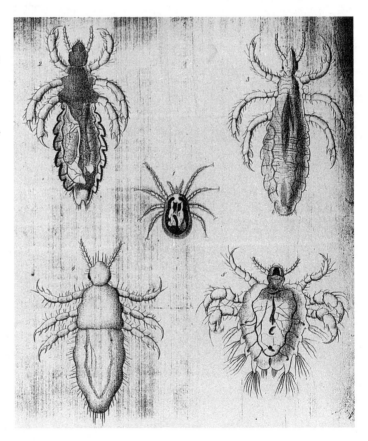

Fig. 1. A mite of some kind (1), a head louse (2), a cloth louse (3), a phthiriasis louse (4), and a crab louse (5). An illustration from Dr. Alt's thesis on phthiriasis, published in 1824. From the author's collection.

tices, the particles of their blood melted together into lice. These speculations met with harsh criticism. The German dermatologist Dr. Kurtz, for example, objected that he had once seen a young pauper woman with large insect-filled boils all over her body; the disease was progressive, and she soon died; the postmortem proved her to be *virgo intacta*, thereby disproving Stegmann's theory that the disease was caused by sexual excesses.

One of the many new cases published in the 1830s and 1840s was that reported by the famous dermatologist Baron Alibert in his *Clinique de l'Hôpital de Saint-Louis*. A young man, swarming with vermin, was taken to this hospital, and there he died "in the most frightful tortures." This case cannot be diagnosed as proper phthiriasis, however, since there is no mention of clusters of insects under the skin. Nor is the drawing of the patient, reproduced by Alibert, consistent with the reliable descriptions of this disease.

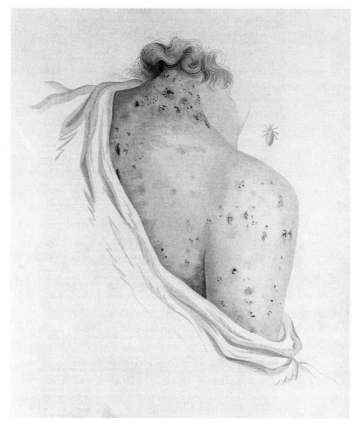

Fig. 2. The only known illustration of a patient with the lousy disease, from Baron Alibert's *Clinique de l'Hôpital de Saint-Louis* (1833). It is doubtful that this individual really had genuine phthiriasis, however; it may well have been a case of generalized pediculosis. From the author's collection.

During this time, the scientists and medical men interested in phthiriasis were divided into several schools. Some of them still accepted Alt's theories about the phthiriasis louse. Others believed that phthiriasis was caused by ordinary body lice. Still others favored the hypothesis that it was caused by some species of mites. Some medical men in Germany and Scandinavia had recognized insects taken from phthiriasis patients as mites rather than lice. A faction of increasing size doubted the existence of phthiriasis altogether, refusing to believe in the existence of such an extraordinary disease if they had not personally witnessed a case.

Several interesting Swedish cases were published about this time, one by the celebrated physician Magnus Huss, who had seen a middle-aged woman with fifty insect-filled tubercles on the chest and loins; these tubercles burst by themselves and the phthiriasis did not recur. The country practitioner Ekman published two cases with all the classical manifestations. Remarkably, in one of them, the patient's neighbors gossiped about the disease being a divine punishment for excessive cleanliness. Ekman gave a very good description of the vermin; they were, he said, white with round bodies and a black dot on the back, and they moved very vigorously. Since lice can hardly be said to move vigorously, the creatures seem more likely to have belonged to some subgroup of mites.

The Vienna Phthiriasis Debate

In 1856 the first serious attack on the legend of the lousy disease was launched by the German practitioner Dr. Husemann of Detmold. He had been called to the hospital in the city of Lippe to examine a soldier who was said to suffer from this most horrible disease. The whole town was buzzing with rumors of the sad fate of the lousy warrior. When Dr. Husemann entered the ward, the man was sitting there, eating soup with a healthy appetite. When asked if he was the soldier with the lousy disease, he answered, "Zu Befehl!" — "At your order, Sir!" A brief examination showed that the man was by no means free from lice, but neither were his brothers in arms; otherwise, he seemed to be in perfect health. Husemann considered the man a mere malingerer and had him restored to his company. Apparently, this odd experience inspired the German doctor to study and review the considerable literature on this ancient disease. He concluded that there was not and never had been any lousy disease. It was absurd that it was still included in many respected dermatological and pathological textbooks, often with a list of some of the historical victims of the disease appended. The legend of phthiriasis as a royal malady lived on well into the nineteenth century. When an alleged case of *morbus pedicularis* was discussed before the Medical Society of London in January 1838, one of those present said that one of the kings of England had fallen victim to this disease, as had, according to report, one of the late royal duchesses. Another famous victim was King Ferdinand of Sicily, an unpopular, reactionary monarch who died in 1859.

Two new cases of the lousy disease were presented by the German country practitioner Dr. Gaulke in 1863. One of them, a vagabond, was taken to Insterburg Hospital full of lice. On his belly and chest were a hundred or so reddish swellings about the size of a pea or a hazelnut. Some of them were open and reached down into the subcutaneous tissues; they contained a mass of living insects, but were completely dry. Gaulke cured the man with petrol baths, but this treatment did not avail with another sufferer of the lousy disease, an old woman, who was "literally devoured by the lice," as in the stories of Sulla and Herod. Dr. Gaulke believed that *Pediculi vestimenti* could lay eggs under the skin, especially in very unclean individuals. He related the high prevalence of the lousy disease in his district to the many Russians, Gypsies, vagabonds, and outlaws residing there. He had calculated that the use of soap in this region was only 1/200 of that in England or France! As a firm believer in phthiriasis, Dr. Gaulke was seconded by the Griefswald anatomist Leonard Landois, who mistook chitinous bars for mandibles and claimed that lice could gnaw a hollow into the skin.

Professor Ferdinand von Hebra in Vienna, one of the foremost dermatologists of the nineteenth century, was fascinated by the riddle of phthiriasis. He circulated word among many of his colleagues that he would be most interested in any cases they might show him. After fifteen years of practice he had seen none, although he had treated about eleven thousand cases of infestation by lice as well as almost every other skin disease. Hebra reacted strongly to the publication of Gaulke's and Landois's papers upholding the ancient myth of phthiriasis. In 1865 his lengthy rebuttal was serialized in the *Wiener medizinische Presse*. Hebra reviewed the literature on phthiriasis and debunked the old cases. He believed that the old stories of lice under the skin were the works of superstition and excited imagination, and he concluded, "Es hat niemals eine Phthiriasis gegeben, noch gibt es heutzutage eine solche" — "There has never been any phthiriasis, nor is there today any such disease." His article provoked heated debate in the Vienna medical journals, and since most participants were prolix, not to say repetitious, writers, it was more than a year before the readers were spared the lengthy serials debating the existence of the lousy disease.

Both Gaulke and Landois took great exception to Hebra's views, and also to his irreverent jokes at the expense of the old victims of phthiriasis and the insults he hurled at latter-day observers of this disease. Gaulke could even present a new case of phthiriasis, a sixty-five-year-old sickly shoemaker, who was taken into Insterburg Hospital swarming with vermin. The man had many small tumors all over the body, from which insects poured incessantly; these swellings itched violently, and the patient sweated excessively. The largest louse tumor was on the back, as had often been the case. It was the size of a walnut, full of insects, and quite dry. Dr. Gaulke managed to cure the man with the same method used in the earlier case; before he was free of vermin, three experienced practitioners were invited to see him, and they all marveled at this strange disease, of which they had never seen the like. When the man was cured, Gaulke had him photographed; it would have been wiser to take the photograph while he was still suffering from the disease, however, and wiser still to preserve some of the insects for examination by an expert entomologist.

Professor Hebra declared himself to be impressed neither by Landois's erroneous arguments nor by Gaulke's odd case report. He was supported by the Danish zoologist Jørgen Christian Schiødte, who had examined the sucker of the cloth louse; he considered it quite impossible that the animal could work its way into the skin with this feeble tool, and stated as his opinion that "the ancient ghost of Phthiriasis could finally be laid to rest among the other dragons and monsters, bred by ignorance."

The Vienna phthiriasis debate resulted in complete victory for Professor Hebra, and the lousy disease was expelled from the pathology books for good. My extensive search turned up only one case, reported by the French surgeon Bertulus in 1870, published after Hebra's final paper. The medical men rapidly changed their minds, and the dermatologist L. D. Bulkley well expressed the new consensus in a paper in the *Archives of Dermatology* in 1881: "All the fabulous stories in regard to lice issuing from abscesses or sores are utterly without scientific foundation, — are, indeed, impossibly absurd." The last thorough review of the legend of the lousy disease occurred in the German *Real-Encyclopedie der gesammten Heilkunde* of 1882; the writer of this section seemed to be quite unconvinced by Professor Hebra's arguments.

Did the Lousy Disease Exist?

During the early years of the twentieth century, the lousy disease was gradually forgotten. No new cases were reported, and in books on classical and biblical medicine, it was mentioned only as an example of the confused medical thinking of those times. The classical historians were obviously puzzled by the havoc wrought by phthiriasis among the eminent figures of antiquity. Some of them advocated the theory that the likeness between the Greek words *phtheir* (louse) and *phtheiro* (to destroy) caused the relatively harmless lice to be blamed for a terrible and ravaging disease with another, unknown etiology; this postconstruction is completely unsupported by the original sources. Others have proposed that the phthiriasis described by Aristotle and others was in fact mere scabies. But although the itch mite lives beneath the epidermis, it does not burrow farther down to live in clusters under the skin; nor does it cause a lethal disease.

Many modern parasitologists have expressed their opinion that the lousy disease never existed and that the belief arose from cases of severe louse infestation on sick and cachectic individuals or myiasis in open wounds. But the legendary disease was to be rehabilitated. In 1940 A. C. Oudemans, emeritus professor of entomology in Arnhem, published a thorough study of phthiriasis and its history. Oudemans was a highly respected acarologist, and his study of *Harpyrhynchus*, a genus of mites, made him inclined to believe that the lousy disease had really existed and that it was caused by a species he called *Harpyrhynchus tabescentium*. Although no modern case of human infestation by this mite has been recorded, there is a good deal of evidence for this hypothesis. Modern descriptions of *Harpyrhynchus* infestation of birds much resemble the classical reports of phthiriasis. These insects are the only mites capable of burrowing under the entire skin, where they live in large clusters of individuals of varying age and size. Harpyrhynchi usually infest birds, and through excessive growth of the mite tumor and the hyperkeratosis they induce, they may even kill their wretched victims. When the tumor, which can be as large as a hazelnut, is opened, thousands of mites of varying size stream out. Microscopic investigation shows that the tumor resembles an encapsulated cyst with septa of collagen; transudation through the capsule nourishes the parasites. Only a small quantity of serous fluid accompanies the innumerable mites when the cyst is opened; no immune reaction or formation of pus occurs — a phenomenon that also puzzled the old phthiriologists. Harpyrhynchi are hardly ever found on other animals, but they have sometimes infested snakes, cats, and squirrels. The length of these alleged killer mites is about 750 μm. By comparison, the head louse is 1–2 mm long and the body louse 3–4 mm; the scab mite, *Sarcoptes scabiei*, is at most 450 μm long. Later acarologists have considered that human *Harpyrhynchus* infestation may well be possible, but no case has ever been observed.

Professor J. R. Busvine, who reviewed Oudemans's hypothesis in his *Insects, Hygiene, and History* (1977), also accepted the idea, but he regretted that no good description or proper drawing of these killer mites existed. His conclusion was that "nothing remotely like the descriptions I have quoted has been actually encountered; so the ancient curse of the lousy disease remains a mystery."

The professor of parasitology at the University of Peking, Reinhard Hoeppli, later published several articles, commenting on Oudemans's theories and describing several remarkable Chinese cases of phthiriasis. Two of these were from a medical work published in 1546, from the time when the lousy Portuguese nobleman Tabora's servants were busy emptying baskets of lice into the sea; the third one was from 1846. Lice living in tumors under the skin seem not to have been unknown to the old Chinese physicians; one patient was cured by an incision of the tumor, but another died in his phthiriasis after the same treatment, with an incessant flow of insects issuing from the aperture. The Chinese doctors also wondered about the etiology of this strange disease, and one of them postulated that impurity and sexual excesses made the tissues of the body transubstantiate into lice.

In Scandinavian folk medicine there are many accounts of a disease called *lusesjukan*, which was very similar to the classical idea of phthiriasis. It was caused by small white insects that were distinct from common body lice; they were said to gather in large swellings or "bags" under the whole skin. A Swedish tale from the early nineteenth century describes the treatment undergone by a man who had large bundles of lice between the flesh and skin. He was seated above a tub of steaming hot water, into which the insects swarmed. The tales emphasized that old and weak individuals were often afflicted with these lice, and the sight of them was considered a portent of impending death. Phthiriasis appears to have been a rather well-known disease in the Swedish and Norwegian countryside. According to a Swedish practitioner of the nineteenth century, "It was often stated by the populace that this person or that had died of phthiriasis, always with the addition that only skin and bones were left when he died." The lore of phthiriasis died hard in Scandinavia, and late nineteenth-century popular books and articles on medicine deplored the credulity of the country people, who hewed to their belief in this "nonexistent" disease.

The last instance of phthiriasis as a royal malady is from late nineteenth-century Sweden. According to local tradition, a famous country quack, the Old Woman of Kungsbacka, successfully treated the king of Sweden for the lousy disease. One version named Oscar I as the lousy monarch, and another

Charles XV. The stories agree that the king had contracted phthiriasis during his foreign travels, and that he distrusted the ability of the royal physicians to treat such a formidable malady. The Old Woman had the king put into the hide of a newly slaughtered ox, through which a multitude of holes had been made to allow the insects to crawl through it to the hairy side of the hide. She had previously cured a clergyman with phthiriasis in the same manner. The king, too, was freed of lice, and he rewarded the Old Woman by granting her the right to practice medicine. This ancient *balneum animale*, or bath in animal flesh, which was mentioned by Paulus Ægineta, was a time-honored Nordic cure for phthiriasis, and was already being recommended in an early eighteenth-century Norwegian manuscript. Thus, the Old Woman of Kungsbacka was well informed how to treat phthiriasis according to old Nordic folk medicine. Her story of the lousy king has been discredited by reliable historians and ethnologists. It only signifies the able self-promotion that helped build the reputation of this popular quack, and also that the concept of phthiriasis as a royal malady still existed in the Swedish countryside.

In all, I found more than ninety alleged cases of phthiriasis in the medical literature. A considerable proportion do not hold up under critical reexamination, if we accept as criteria that the insects must be said to have caused swellings or tumors by occurring in clusters under the skin. Especially in the early nineteenth century, there was considerable confusion in the nosology and nomenclature of the louse-related diseases. In English the names phthiriasis and *morbus pedicularis* were used interchangeably for both the lousy disease and ordinary lousiness. In French *phthiriase* was a similarly unprecise term, while *maladie pédiculaire* was generally used for genuine phthiriasis; the influential Baron Alibert brought further confusion by introducing the new denominator *prurigo pédiculaire* from Willan's nosological system. In German, *Läusesucht* and *Phthiriasis* were synonymous, and both were also used for common lousiness. This slackness in terminology led to inevitable confusion: it was not rare that cases of scabies or common louse infestation were published under the name of phthiriasis, often with the conclusion that the patient had been very dangerously ill of this "most horrible of diseases" but had been saved by the vigilance and skill of the physician. It should also be noted that infestations with common body lice could become extremely severe in unhealthy, aging, and malnourished individuals of lax hygienic habits. In such cases, there would be a general skin eruption with secretion and encrustation; wounds and scratching would lead to secondary infection; and the louse infestation would be almost impossible to treat with the contemporary antiparasitic regimes. Several patients were reported to die of this kind of "lousy disease"; the insects were usually recognized as common body lice, and they did not occur in clusters under the skin. Skeptics used such cases to prove that the only remnant of the legendary disease was this form of severe louse infestation.

A thorough search of the literature reveals forty-two cases of classical phthiriasis reported between 1540 and 1870. In nine of them, the vermin were recognized as mites, and in several of the others, they were said to be unlike ordinary lice, being small, white, and very agile. Some observers also described a black dot on their backs, corresponding to the *Harpyrhynchus* mite's dorsal shield. Some historians conjecture that phthiriasis was much more common during classical antiquity. It is true that the frequent mention of the disease in medical and other literature seems to point in this direction, but the widespread legend of phthiriasis as a divine punishment makes it difficult to assess what role it really played during this time. Phthiriasis was often mentioned during the sixteenth and seventeenth centuries, but for some reason, the number of classical cases decreased during the 1700s. In contrast, no fewer than twenty-four well-described cases were published between 1813 and 1870; they help to explain why this fabulous disease, although made "impossible" when the theory of spontaneous generation of insects had been disproved, remained within the pathology books. Phthiriasis was most common in Germany, Scandinavia, and France; only a few reliable British cases were ever reported and not a single American one. The lousy disease was by no means unknown to German dermatologists and general practitioners in the early and middle 1800s, and when this disease was discussed before the Swedish Society of Medicine in 1849, several of those present claimed to have encountered the lousy disease themselves, without publishing the cases. Since phthiriasis was most common in poor and underdeveloped areas of Europe, there were probably several unpublished cases for each published one; the doctors who encountered the disease were not university professors but hardworking rural practitioners, who often did not have the time or skill to write case reports and get them published.

Although few would miss such a revolting malady, it remains a mystery why phthiriasis disappeared so suddenly in the 1870s. Advances in hygiene and general living conditions probably played a part, but during its heyday phthiriasis also claimed victims among the wealthy, young, and well nourished. Perhaps the species of mite causing the disease died out at about this time. We must doubt, however, whether the disease was really so formidable. Several nineteenth-century textbooks pronounced it incurable, but this highly pessimistic view of the natural course of phthiriasis, which had its origin in the legend that it was a divine punishment, did not reflect reality. Twenty-two of the forty-two patients were in fact completely cured. In one case, the disease even went into spontaneous remission, and in others, treatment with petrol baths, mercurial ointment, sulphuric baths, or other regimes were successful. The prognosis was much better when there was only one louse tumor, and all but one of the eight patients with this form of the disease were cured. It is interesting to speculate whether different species of mites might have caused these two forms of the disease.

There are forty-two similar case reports describing a disease resembling the one reported in classical times, some of them written by the foremost physicians of the seventeenth and eighteenth centuries. After rereading ten to fifteen of the best-described cases, such as those by Valentin, Lindestolpe, and Gaulke, it is impossible to reach any other conclusion than that a disease of these characteristics really existed at the time. It is a strong argument in favor of the existence of phthiriasis that descriptions of a similar disease can be found in both traditional Chinese medicine and Nordic folk medicine. It seems likely that the disease was caused by some tumor-forming mite species such as *Harpyrhynchus*, but we cannot prove this theory with certainty in the absence of a good illustration of the parasites.

Thus, the tyrants, fornicators, and enemies of religion of the present day should beware: the last of these redoubtable killer mites may not have breathed its last, and the ancient curse of the lousy disease may not have been exorcised. Transgressors would do well to tremble and cover their heads in fear, asking themselves, like the furies in Josiah Sylvester's translation of Guillaume du Bartas's famous poem *La Semaine*: "But with the griefs that charge our outward places / Shall I account the loathsome *phthiriasis*?"

Giants in
the Earth

> *Then, too, the claims of the monster Polyphemus
> must also be borne in mind. His skeleton had
> been dug up at Trapani in Sicily in the four-
> teenth century, so it was said, and showed
> him to have been some three hundred feet in
> height — there was a real giant for you!*
>
> Sir Osbert Sitwell, *Tales My Father Taught Me* (1962)

IN 1577 A HUGE SKELETON WAS FOUND AT AN
excavation near the city of Lucerne in Switzerland. No one knew the
nature of these gigantic remains, and Professor Felix Plater, the famous
Basel anatomist, was called in to examine them. After scrutinizing the bones,
he told the burgomaster and the magistrates that they were the remains of a
nineteen-foot antediluvian giant. The giant of Lucerne became the pride of
the town, and his bones were put on a permanent *lit de parade* in the city hall;
they were admired by the townsmen and visitors alike. In 1786 the celebrated German naturalist Johann Friedrich Blumenbach took the opportunity to examine the giant's skeleton while he was passing through Lucerne.
To the disappointment of the burgomaster, he declared that the skeleton had
once belonged to a mammoth.

From antiquity until the early nineteenth century, a great many of these giants in the earth were discovered all over the world. Some had body parts more than a hundred times as large as those of a normal human being. The colossus of them all was the monstrous Polyphemus, whose skeleton was found in Sicily in the fourteenth century. It was believed that the human race had steadily decreased in size and that before the Flood people had been at least fifty feet tall. The church was instrumental in spreading the doctrine of the antediluvian giants, which had support from the Old Testament. Not a few theological theses were dedicated to this subject, sometimes with the most startling speculation about the biblical giants' origins and way of life. To persuade the skeptical, clerics hung the giants' bones in the churches for the inspection of all and sundry, and these bizarre relics could be seen in several European cathedrals.

The Giants of Legend

In olden times there was no lack of giants; the mythologies of most civilized peoples contain tales of gigantic men who waged war, usually with notable lack of success, against the ancient gods and heroes. In Greek mythology, the forty-five Titans were the children of Uranus and his own mother Terra; they were all of enormous size and strength. When they castrated their father with a sharp scythe, the flowing blood engendered even more horrible creatures: the Gigantes, men of prodigious size and hideous aspect. When the Titans had been defeated after a failed uprising against Zeus, their monstrous relatives came to the rescue: the Gigantes attacked the gods armed with huge rocks and tree trunks; they scaled the heavens by heaping Mount Ossa upon Pelion. One of the Gigantes, the fearful Typhon, could breathe fire and smoke through the eyes and mouths of his one hundred heads; when he appeared outside the walls of Olympus, all the gods were terror struck, and they hid in Egypt in the guise of various animals. After recovering from the shock, the gods employed Hercules to fight the Gigantes, and the monstrous creatures were all killed in appropriate ways: some were crushed under mountains, others flayed alive, others beaten to death with clubs, and still others thrust head first into the craters of volcanoes.

One of the most enigmatic passages in the Old Testament is Genesis 6:4, "There were giants in the earth in those days; and also after that, when the sons of God came in unto the daughters of men, and they bare children to them, the same became mighty men, which were of old, men of renown." Speculation was rife about the nature of these *filii Dei*, or "sons of God." According to the noncanonical Book of Enoch, certain angels sent by God to guard the earth were very much impressed by the beauty of the terrestrial women, with whom they bred a race of evil, gigantic, semihuman beings. In Milton's *Paradise Lost*, Satan tells Belial:

> *Before the Flood thou with thy lusty crew,*
> *False-titled sons of God, roaming the earth,*
> *Cast wanton eyes on the daughters of men,*
> *And coupled with them, and begot a race.*

The giants walked the earth until the Flood purged the world of them. According to some medieval theologians, the only purpose of the Deluge was to punish these monstrous creatures; God considered it worthwhile to destroy all creation to get rid of them. This high price seems even more immoderate when we consider that the Old Testament also mentions the existence of several postdiluvian giants. Moses smote many of them until, at the end, only one of the giants of old remained, King Og of Bashan. He was said to be three thousand years old, and if some early Jewish theologians are to be believed, his size was truly enormous. King Og had survived the Deluge by wading alongside the Ark; the waters came only to his knees. When he got hungry, he speared a whale-fish and roasted it near the sun. According to another version, King Og had made a deal with Noah, who let the giant hitch a ride, sitting outside on the Ark; in return, the king was to serve Noah and his descendants as a slave for the rest of his life. King Og apparently did not honor this treaty after Noah's death, since the Old Testament mentions him as a powerful enemy of the Israelites in Canaan. The writer of Deuteronomy declares that the king's iron bedstead, which the children of Ammon kept at Rabbath, was nine cubits long. According to one of the Targums, King Og was much taller than that, and his feats much stranger. Once, he tore up a high mountain from the earth and carried it off, planning to throw it on the camp of the Israelites. The Lord let a worm undermine the mountain, however, making it fall on his shoulders; at the same time, the king's teeth grew out in all directions, preventing him from casting it off his head. When Moses saw the colossus in this unique predicament, which is probably unequaled even in the bloody annals of divine retribution, he struck King Og on the ankle bone with his great axe; the giant fell heavily to the ground and never rose again. His thigh bone was twelve leagues long, and Moses ordered his men to make it into a bridge to span a broad river. Another biblical giant, the famous Goliath of Gath, belonged to another evil gigantic tribe in western Palestine. According to the First Book of Samuel, he was "six cubits and a span" (nine feet, nine inches) tall, and thus a mere shrimp beside King Og.

In ancient Greece and Rome, many fabulous tales were told about the huge size of the legendary gods and heroes. It also seems to have been a common belief that the bodily size of the human race had declined steadily with time. Homer regretted that the champions and demigods of old were but memories and that the weak and feeble frames of his contemporaries did not equip them for any heroic deeds. Virgil agreed with these sentiments; he wrote that when the skeletal remains from the Roman civil war were discovered, they were to be revered for their huge size. The actors who played the parts of gods or heroes in historical plays and tragedies used to increase their size by wearing buskins that almost resembled stilts, as well as gauntlets to lengthen their arms and cushioned robes to increase their girth.

According to Pliny, when a mountain in Crete was split by an earthquake, a human skeleton, 46 cubits (77 feet) tall, was found standing upright in its midst. Some learned Romans presumed it to be the grave of the legendary giant Orion. In the same island, another huge skeleton was found after a flood; the consul Metellus and the legate Lucius Flaccus, who saw it, accepted it as the skeleton of one of the giants of old. Several other Latin writers describe similar bizarre findings, among them Philostratus in his book on the heroes. According to Suetonius, Augustus Caesar preferred to decorate his house with antiquities and natural curiosities, instead of filling it with statues and works of art like many wealthy Romans; in his collections, the emperor kept several giants' bones.

Among the early Jewish and Christian theologians, the common opinion was that a race of giants had been bred through the scandalous *mésalliance* between the *filii Dei* and the wanton daughters of the earth, and that they were inherently evil. In *The City of God*, Saint Augustine disagreed with these notions, which did not fit his theological ideas. He interpreted the passage in Genesis to mean that there were already giants in the earth when the sons of God took wives among the daughters of men; thus, the giants were just extremely large people. All the antediluvians were much taller than the puny races that had taken over after the Flood, and the giants were merely the biggest of them all. Saint Augustine may have arrived at this conclusion after attending the exhibition of a giant's molar, which was more than a hundred times heavier than one of his own.

76   Polyphemus Resurrected

In the year 758, a gigantic skeleton was found at Totu in Bohemia. The giant's head could only just be encompassed by the arms of two men joined together. The shin bones, kept in the collection of a medieval castle as late as 1764, were 26 feet long, by which it was conjectured that the giant's entire body had exceeded 110 feet in length. The most enormous of all giants in the earth, more than 330 feet long, was found in the year 1342 near the mountain of Eripana in Sicily. It seemed to be in a sitting position, with one hand resting on a club as long as a ship's mast. The bones crumbled when touched, but the teeth and the skull were taken to a church in Trapani, where they were shown for many years. The shape of the skull seemed to indicate that the giant had had only one eye, and the gigantologists soon agreed that the bones were those of the far-famed cyclops Polyphemus, described by Homer as

> *A form enormous; far unlike the race*
> *Of human birth, in stature or in face;*
> *As some lone mountain's monstrous growth he stood,*
> *Crown'd with rough thickets and a nodding wood.*

Many medieval theologians were busy pondering the role of the giants in biblical history. It was difficult to understand, for example, why, if the giants were really the sons of angels, they were evil enemies of religion. Since it was also commonly believed that all antediluvians had been very large, then what was the difference between the good patriarchs and the evil giants? Some theologians postulated that although the early humans had been truly enormous, the giants were even bigger; furthermore, they were distinctly subhuman, in spite of their semiangelic parentage, and lacked the power of higher reasoning. It was also an enigma to the early theologians and philosophers why, if the Deluge had been intended to kill all the giants, some of them, such as King Og and Goliath, had reappeared after the Flood. Since the Bible directly states that only Noah and seven people survived on board the Ark, there was much speculation about which of them had been the giant.

In 1498 the Italian monk Annius of Viterbo announced a remarkable discovery. He claimed that the historian and astrologer Berosus, active in the third century B.C., had been the librarian of the Babylonian temple library and that he had written a historical chronicle by the use of many forgotten books on its shelves, some of which, Annius claimed, had belonged to Noah himself before being stolen by Nimrod when he went away to found Babel.

These books contained the chronicles handed down from Adam to Noah, and when the library was destroyed, the chronicle of Berosus was the only version that remained; it had itself been lost for centuries until Annius had been able to procure a copy from an Armenian priest. In reality, Annius had forged the entire work, using it to put forth his own eccentric interpretation of the Bible and the works of Flavius Josephus. The counterfeiting monk declared that Noah and all his family had been giants. This bold stroke neatly solved the problem of the giants' survival after the Deluge. Annius also constructed an elaborate family tree for Noah and his descendants, the first of all royal families, all of them giants, but pious, gentle, and good. They had founded a pre-Roman empire of vast dimensions.

The original Latin edition of Annius was an obscure and academic work, but it nevertheless achieved considerable popularity. Several editions were published in sixteenth-century France, and a certain Jean Lemaire de Belges provided a popularized French version of the gigantology of Annius, which was quite a best seller. Many early French historians were greatly influenced by these marvelous stories, and the obscure fictions of Annius reappear in many scholarly tomes, merged with preexisting local traditions of giants and their battles with humankind. Patriotic historians took great pains to discover or invent legends of giants connected with the founding of national states and capitals, for these linked the nation to the ancient empire of Noah's descendants. Discoveries of giants' bones were also of prime importance to the epigones of Lemaire, for they provided evidence that the territory had once been the home of the ancient giants. It was particularly important that the giant be identified, and French historians pored over the Bible and the works of Annius and Lemaire to discover the identity of the colossus who had once walked the earth of their native country.

In the sixteenth and seventeenth centuries, traditional biblical gigantology coexisted with the startling new ideas of Annius of Viterbo and Jean Lemaire. Few people doubted the existence of antediluvian giants, and if any lacked confidence in the official gigantology, the churches were kept well stocked with giants' bones to convince them.

The first man of science to oppose this dogma directly was the Dutch doctor and antiquary Johan van Gorp. While researching an antiquarian work on the foundation of Antwerp, he had studied a famous giant's tooth, which was the most prized exhibit in the city museum. According to local tradition, this giant had been vanquished by Julius Caesar's son Brabo, and his defeat had made available the ground for the building of the castle and city of Antwerp. A month earlier, van Gorp had seen two fossil elephant skeletons, complete with tusks and grinders, which had been discovered by workmen digging a canal. He compared the giant's tooth with the elephant's grinder, and they matched perfectly. This unexpected discovery made him distrust all stories of legendary giants. In his book *La gigantomachie*, which formed the second part of his *Origines antwerpianae*, published in 1599, van Gorp bluntly declared that all the stories of giants in the earth were mere fables and that the height of antediluvian man had not exceeded that of his own contemporaries.

Wary of the church's great influence and careful to avoid charges of heresy, van Gorp maintained that the biblical giants had been "giants" in other ways than bodily size, for instance, in their truly gigantic appetite for war and destruction. Nevertheless, the publication of *La gigantomachie* raised considerable furor in theological circles. In 1580 the Frenchman Jean Chassagnon wrote a treatise, *De gigantibus*, solely to refute van Gorp's dissenting arguments, but he had little new to add and merely repeated the same fabulous old tales. One of his most powerful arguments was that van Gorp had insulted God's creative power; it had been the Almighty's intent to create giants as well as dwarfs to adorn the universe. Thereafter, for more than one hundred years, it remained a useful thesis project for junior theologians to try to refute the heretical van Gorp.

A Gigantic Scandal in France

The most famous of all giants in the earth was discovered on January 11, 1613, in the grounds of the ruined castle of the marquis de Langon, near St. Antoine in the French province Dauphiné. Some masons digging in a field known since ancient times as "Le Terroir du Géant," or "the giant's ground," came upon an elaborate brick tomb, thirty feet long and twelve feet wide. The masons opened the tomb and found within it a complete human skeleton, more than twenty-five feet long and ten feet wide across the shoulders. The giant's teeth were the size of an ox's hoof. The local surgeon, M. Mazuyer, and the village clerk, David Bertrand, went to examine the remains. Several ornaments and urns were also found, as well as a gray stone block with the inscription "Teutobochus Rex." The erudite surgeon realized that the remains must be those of King Teutobochus of Cimbria, a legendary giant who was defeated by a Roman army under Marius and taken to Rome as a prisoner of war. In the triumphal procession, the giant's head had towered over the trophies carried atop the long spears of the legionnaires.

M. Mazuyer was impressed by the considerable interest in the skeleton among the local worthies. Clergymen, nobles, and antiquaries bared their heads at the giant's *lit de parade*. The sight of these solemn ceremonies gave him a brilliant idea. After bribing the local collectors and noblemen with some ornaments, teeth, and medallions to make them give up their rights to the skeleton, he set out for Paris at the head of several horse carts loaded with bones, ornaments, and inscribed stone slabs from the grave. The clever M. Mazuyer had written to several friends in Paris beforehand, asking them to spread the word about the forthcoming state visit of King Teutobochus.

The result must have exceeded the resourceful village surgeon's wildest dreams. When M. Mazuyer and his principal assistant, the village clerk Bertrand, reached Paris on July 20, they were officially welcomed by the mayor and the superintendent of the royal museum of medallions and antiquities. The marquis de Langon, on whose grounds the skeleton had been found, invited Mazuyer and Bertrand to stay at his house, where they hobnobbed with dukes and counts; none of these magnates left the exhibition of the skeleton without giving Mazuyer a considerable fee. The high-water mark of his career was reached when King Louis XIII ordered the skeleton of his royal counterpart brought to Fontainebleau and placed in the queen's chamber. There it was seen by the entire court. The king was most impressed by the huge bones, and he asked Mazuyer if it were true that giants of this size had once walked French ground. The delighted surgeon assured him that they had. A courtier added that these huge men would make a powerful army, but the king replied that they would also eat the country clean of food in no time!

Mazuyer had become one of the celebrities of Paris. In order to keep his position in the limelight, he let the Jesuit Father de Tournon write a pamphlet titled *Histoire véritable du géant Theutobochus, roy des Teutons*, which became hugely popular among the Parisians; three editions were published in the first year alone. A friend of Mazuyer's, the surgeon Nicolas Habicot, who was known for his quarrelsome disposition, shared his surgical colleague's interest in gigantology. Habicot took Mazuyer's sensational finding as incontrovertible proof that giants had once walked the earth. In his book *Gigantostéologie*, dedicated to the king, he put forth this opinion in the most belligerent fashion. The famous anatomist Jean Riolan, who doubted that the skeleton really was that of a man, took issue with his views, thus beginning one of the most acrimonious quarrels in French medical history. Both men were indefatigable pamphleteers, and there are at least seventeen of their scurrilous and often amusing publications in the Bibliothèque Nationale. The quality of the debate declined continuously as pamphlets were exchanged. The first of them were reasonably well-argued summaries of the arguments for and against the existence of antediluvian giants, but the later pamphlets contained abusive diatribes against the opponent's professional honor, personality, and parentage. Habicot was worse, but Riolan also demonstrated considerable skill in the art of vituperation. He was apparently of the firm opinion that not only Habicot and Mazuyer but all French surgeons were little better than low-born horse doctors, ignorant of anatomy and osteology. Unwisely, he even put in some satirical remarks about that famous doyen of French surgery Ambroise Paré. Several other surgeons went over to Habicot's side, and according to one publication, Riolan was forced to cower behind the shuttered panes of his Paris house while gangs of surgical apprentices bayed for his blood on the pavement outside, eager to restore the honor of Paré, Habicot, and King Teutobochus.

After the quarrel between Habicot and Riolan had developed into a full-blown feud, and all the innumerable students and surgical apprentices of Paris took sides, the exhibition of King Teutobochus was in danger of becoming a battleground. Once, a gang of drunken students even tried to kidnap the giant, but fortunately, Mazuyer had had the foresight to employ a party of toughs to guard the king. They managed to repulse the students, not without bloodshed. The former village surgeon was not a man of violence, and late in 1614, he left Paris to take the skeleton on a tour of Germany and Flanders. Later the same year, Riolan managed to gain the upper hand in the controversy. He had sent a spy to Dauphiné to interrogate some of the masons who had found the giant, and the blabbering workmen gave away the secret: M. Mazuyer had made several grave ornaments and inscriptions with his own hands! Furthermore, Riolan had himself examined the king's skeleton, and he was entirely convinced that it had once belonged to an elephant. This did not end the controversy, since the furious Nicolas Habicot abandoned theoretical arguments in favor of fresh torrents of abuse. He called Riolan, who had written a treatise on hermaphrodites, "Le Docteur Hermafrodite" and also published a highly insulting Latin anagram on his name: "Joannes Riolanus, en Laurus in asino, en asinus in lauro, sine lauro inane," meaning that Riolan was only a jackass decorated with academic laurels!

Although the final pamphlets in this gigantic quarrel were not written until 1618, most people were by this time convinced that Riolan was right. Habicot's unbalanced acrimony had further damaged his reputation. The outcome of the debate also affected Mazuyer's prospects. When his European tour brought him to Marseilles, he found the public's interest in gigantology at a low ebb. Furthermore, Molière's troupe of actors was playing in the house next to Mazuyer's, and it nettled the ex-surgeon, who had once presented King Teutobochus at court, to hear passersby cracking jokes about the giant and his fraudulent impresario while walking toward Molière's evening performance. After a couple of weeks, Mazuyer had to declare bankruptcy. His only possession of value, the giant's skeleton, was seized by his landlord in lieu of several weeks' rent. The disillusioned Mazuyer returned to his native village to resume his practice. It is to be hoped that his patients, who had been without a surgeon for five years, did not tease him with unkind hints about his part in this gigantic swindle. The Marseilles landlord appears to have had some respect for the remains of King Teutobochus, since he did not sell them to some rag-and-bone merchant; instead, they were stored in the attic of his own house. More than two hundred years later, a French naturalist discovered them there, and they were transported back to Paris in 1832. Here, a certain M. Jouannet examined them, finding that neither Habicot nor Riolan had been right; the bones were those of neither a giant nor an elephant but a mastodon.

The Heyday of Gigantology

A particularly thorough review of mid-seventeenth-century gigantology can be found in Father Athanasius Kircher's impressive *Mundus subterraneus*, published in 1664. In a time of polymaths, Father Kircher was one of the most erudite; he was also a copious writer, producing folio volumes, mostly compilations, in most of the current branches of natural science, including a three-volume treatise on the structure of Noah's Ark. As its title hints, *Mundus subterraneus* deals with everything found under ground. It contains several interesting speculations about the inner structure of the earth and the connection between active volcanoes and the center of the planet. But however advanced he was as a geologist, Father Kircher remained a firm adherent of the traditional fables about giants in the earth. In a remarkable figure, he illustrated the huge size of the antediluvian giants: a "Homo Ordinarius" is a mere insect, and even Goliath could have been trampled underfoot by the largest giant, the famous Polyphemus whose skeleton was unearthed in Sicily in 1342; he is shown with two eyes, although he was believed to have been a cyclops. The "Helvetius Gygas," who is about twice the size of Goliath, is our old acquaintance from Lucerne. These gigantic bodies were reconstructed by anthropometric calculus from one of the "giant's bones" that resembled a corresponding one from the human skeleton. An elephant's grinder could weigh two hundred times more than a human tooth, and the

82   giant's body acquired a similar magnitude. Apparently awed by the size of these giants, Kircher hoped they would inspire the reader of his book with reverence for the works of the Almighty. A reader with an independent mind might wonder instead how these giants would get their nourishment. The monstrous creature would kick an elephant out of his way like a tennis ball, since it would not even fit him for an hors d'oeuvre, before eating a shoal of whales one by one, like sardines, and drinking half the Danube.

In England there was a long and illustrious tradition of gigantology. According to the fourteenth-century chronicles of Ralph of Coggeshall, two teeth of a giant "of such a prodigious Bigness, that Two Hundred of such Teeth, as men ordinarily have now, might be cutt out of One of them" were

Fig. 1. Giants in the earth, from *Mundus subterraneus* by Athanasius Kircher, published in 1664. From the author's collection.

shown in a village called Edulfiness. During the 1600s, gigantic remains
were frequently exhibited in London. Turner the naturalist saw a gigantic
thigh bone in 1610, and another one was to be seen at a private museum near
the Mitre, according to an advertisement in the *News* of 1664. One of the
rarities of John Tradescant's museum in Lambeth was the huge bone of a
giant. When the English diarist John Evelyn went on a journey to France
and Italy in 1644–1645, he took every opportunity to see natural curiosities
in various museums and private collections. In a chapel in the French city of
Bourges he saw "the bones of one Briat a Gyant of 15 cubits-high," which
had been kept there since 1456. In Rome he visited the cabinet of curiosities of a certain Cavaliere Gualdi, which contained the thigh bone of a giant, whose authenticity was affirmed by reliable anatomists. In a church in
Rome Evelyn saw other "huge bones, which they affirme to have ben of some
Gyant." James Paris du Plessis, an author on monstrosities whose work is
still kept in the manuscript collections of the British Library, was much more
skeptical. In the church of the Jacobites at Valence, he had seen several
bones, which had allegedly belonged to a famous, twenty-two foot giant
named Bruard. Although the giant's portrait was also on exhibition, James
Paris du Plessis insisted that "them Bones are Rather the Bones of an Elephant than those of a Man."

The antiquary Robert Plot was another enthusiastic gigantologist. He
described a skeleton found in Cornwall, nearly nine and a half feet tall, in his
Oxfordshire. He had read the account of the giant Gabbaras, brought by the
emperor Claudius from Arabia to Rome; this giant was the tallest man seen
in his time, nine feet, four inches in height. Solely on the evidence that the
proportions of Gabbaras fitted those of the Cornwall skeleton, Dr. Plot presumed that Claudius had brought the giant with him to Britain and that
Gabbaras had died there.

Sir Thomas Molyneux, the distinguished Irish physician and naturalist,
was much more skeptical. In a long article on giants in the *Philosophical Transactions* of the Royal Society in 1700, he agreed with van Gorp that not a few
of the relics of giants in the earth were in fact "Bones belonging to some of
the biggest Quadrupeds, as Elephants, or some of the largest sort of Fishes
of the Whale-kind." Nevertheless, he was not prepared to discard traditional
gigantology altogether, for he had examined a huge frontal bone belonging
to the medical faculty of Leiden, which was undoubtedly human in origin.
Furthermore, he drew a parallel to the animal kingdom; if a Shetland pony
could belong to the same species as a Northampton draft horse and an "Irish
Woolf-Dog" was closely related to a puny lap dog, it was likely that similar

84   "Gigantick Breeds" had once existed within the human race. Sir Thomas
Molyneux seems to have been right about the human origin of the Leiden
frontal bone, but an alternative explanation may be that it had belonged to a
hydrocephalic individual.

In the early 1700s the North American continent also made a notable
contribution to gigantology, through a letter from the Reverend Cotton
Mather to the secretary of the Royal Society of London. Mather, who was a
well-known scholar and antiquary, sent many communications to this famous body on subjects ranging from descriptions of American plants and
rattlesnakes to apparitions, monstrosities, and medical prescriptions revealed
in dreams. One of the letters described several teeth and bones from an
American giant discovered at Claverack, near Albany, New York, in 1705.
One of the giant's teeth had already been sent to the Royal Society by Lord
Cornbury, governor of the province of New York.

Cotton Mather was an old-fashioned gigantologist in the tradition of
Athanasius Kircher, bombastically proclaiming that antediluvian giants had
once walked American soil and taking the discovery as "an Illustrious Con-
firmation of the *Mosaic History*, and an admirable obturation on the mouth of
Atheism!" Mather had apparently spent quite some time studying gigantology, since he quoted many of the older authorities on this subject; although
"the Credit of *Pliny*'s Relations runs pretty Low among the Learned in our
Dayes," he approved of the majority of the others. He regretted that no copy
of Jean Chassagnon's 1580 thesis had crossed the Atlantic, "but what matters
it," he consoled himself, "as long as the *Giants* themselves have come over to
America!" Some American skeptics had advanced the opinion of Johan van
Gorp, that the bones and teeth were those of an animal and that the biblical
giants were but metaphorical ones, but Mather airily dismissed these no-

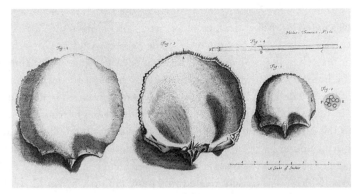

Fig. 2. The huge fron-
tal bone of Leiden,
from the article by Sir
Thomas Molyneux in
the *Philosophical Trans-
actions* of 1700. From
the author's collection.

tions. He was audacious enough to state that he did not think much of van Gorp's treatise. Cotton Mather was seconded by a clerical colleague, the Reverend Edward Taylor, who wrote a lengthy poem about the giant, containing the lines: "This Gyants bulk propounded to our Eyes / Reason lays down nigh t'seventy foot did rise / In height."

One of the most astounding theories about giants in the earth was formulated by the French academian M. Henrion, in a lecture before the Academie des Inscriptions et Belles-Lettres in 1718. After months of complicated calculations, based on figures from the Bible, the Talmud, and the works of some early Latin writers, he had constructed a table of the dramatic decline in human stature since the Creation. Adam had been a 134-foot colossus, and Eve a fitting spouse at 128 feet. Since Noah had been 111 feet in height, the 30-foot Abraham and the 14-foot Moses would have looked like midgets next to him, to say nothing of the puny Hercules, who had been a mere 11 feet in height. M. Henrion's estimate of Adam's stature was quite moderate compared with those of certain early rabbinical writers, who claimed that Adam's head had penetrated the skies when he stood upright and that he could simultaneously touch the North Pole with one hand and the South Pole with the other. The angels were very much frightened when they saw this monster, and they persuaded God to shrink him to a more appropriate size, 330 feet in height; he remained of this stature until he ate the forbidden fruit. M. Henrion regarded the inevitable shrinking of humanity as a divine punishment and believed if it had not been slowed by the steady progress of Christianity among the heathens, his contemporaries would have been as small as fleas, and the entire French Academy could have been in session inside his snuffbox.

Sir Hans Sloane as a Giant Killer

Sir Hans Sloane, Baronet, was a wealthy London physician and botanist who had the wide range of interests typical of the early seventeenth-century natural scientist. For thirty-two years, he was the secretary of the Royal Society of London and the prime mover of this famous body during an important phase of its development. Sir Hans Sloane's enormous collections of antiquities and natural history specimens were the foundation of the British Museum. In 1728 when Sir Hans was sixty-six years old and at the height of his distinguished career, he wrote a paper on the fossil bones of elephants, which was published in the Royal Society's *Philosophical Transactions*. This important paper also contained a critical review of the many purported

86   discoveries of giants' skeletons all over Europe during the previous two
hundred years. In the 1720s old-fashioned gigantology according to Saint
Augustine and Annius of Viterbo was still very much alive. Sloane's investigation had probably been inspired by a thesis titled *Homo diluvii testis*, by the
Swiss naturalist J. J. Scheuchzer, which was published in 1726 and ab-
stracted at one of the Royal Society's meetings. In 1727 the *Philosophical
Transactions* had contained a secondhand account by Dr. Simon Degg, FRS,
of a nine-foot human skeleton discovered in Derbyshire. Furthermore, Sir
Hans Sloane himself had seen the fore fin of a whale exhibited in London as
a giant's hand, and a huge lumbar vertebra sent to him from Oxfordshire had
also turned out to be of cetacean origin.

Sir Hans Sloane's knowledge of comparative anatomy was in advance of
his time. Reexamination of many of the old stories of giants in the earth
convinced him that these fossil bones belonged to whales, elephants, and
mastodons. In several instances, the seventeenth-century gigantologists had
illustrated their books with figures of the teeth and bones, and Sloane was
able to recognize several purported giants' teeth as elephants' grinders, including a famous specimen found by the Swedish army outside Krembs in
1645. Sloane thought even the large skeleton of Polyphemus discovered in
1342 was "not unlikely to have been the Skeleton of a large Elephant." Indeed, the whole concept of cyclopean giants may be the result of misin-

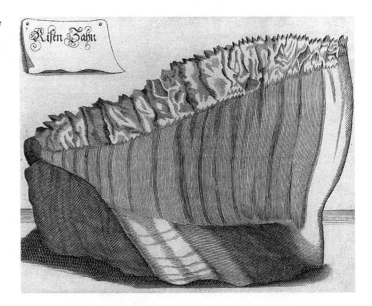

Fig. 3. The tooth of the
"giant" found by the
Swedish army outside
Krembs, from Sir
Hans Sloane's article
in the *Philosophical
Transactions* of 1726.
From the author's
collection.

terpreting elephants' skulls. Their huge merged nasal openings resemble a single, central eye socket.

The Danish naturalist Ole Worm had been an enthusiastic gigantologist, describing and figuring several gigantic remains in his *Museum Wormianum*. Sloane identified a giant's tooth, dug up near Aalborg, as that of an elephant rather than a Danish Polyphemus. Sir Hans also entertained suspicions about bones that were prized exhibits in the private museum of the king of Denmark, but in the absence of an illustration, he was unable to classify them. These fossil bones are still kept at the Museum of Zoology in Copenhagen, and they have been identified without any doubt: one of the Danish "giants" was once a whale, and the other a mammoth. Similarly, an alleged giant's tooth kept at the Musée d'Histoire Naturelle in Bordeaux has been demoted to an inferior molar of an elephant.

In 1678 Emperor Leopold I of Austria was invited to make a purchase that illustrates the economic value of the gigantic bones and teeth. Some Levantine merchants had sent a giant's tooth from Constantinople to Vienna, to be offered for sale to the emperor. They enclosed a certificate of authenticity, signed by several learned antiquaries, and the story of how the tooth had been discovered. In a deep subterranean cavern near Jerusalem, intrepid explorers had found a grave with the Chaldaic inscription "Here lies the Giant OG." Although it seemed that they had found the resting place of the far-famed king of Basan, the explorers did not hesitate to rifle the tomb for whatever bones and teeth they could carry with them. The merchants had had King Og's tooth valued at ten thousand rixdollars, but out of reverence for the emperor, they offered it to him for a mere two thousand. When the imperial antiquaries were consulted, they dissuaded the emperor from purchasing the tooth. They could not recommend the expenditure from the imperial treasury in such highly suspicious circumstances, especially since the emperor already had two giant's teeth in his collection. King Og's tooth was sent back to the wily Levantine merchants, to their great chagrin. Fifty years later Sir Hans Sloane was able to demonstrate that this valuable tooth, as well as the two others in the imperial museum, had once belonged to elephants.

In seventeenth- and eighteenth-century Scandinavia, gigantology followed the same rather contradictory pattern as in the rest of Europe. Giants' bones were exhibited in the churches as evidence of the literal truth of the Old Testament, and at the same time, patriotic antiquaries revered them as evidence that a tribe of giants had once inhabited the Nordic isles. In 1705 a giant's skeleton was unearthed in Norra Vånga, a village in central Sweden,

and taken to the Cathedral of Skara. After five years, the bones were removed to the Nosocomium Museum in Upsala, at the order of Professors
Olaus Rudbeckius and Lars Roberg, who were both keen gigantologists.
When a young Upsala student, Anders Unge, who was born in Skara, was
to become a doctor of philosophy, his friend, the twenty-two-year-old Eman-
uel Swedenborg, wrote a beautiful Latin poem in his honor, alluding to the
giant:

> *Sunt Gothia nuper spatiosi membra gigantis*
> *avecta, ast cerebro, ast ingenioque carent.*
> *Fertilis haes tellus alium nunc mittit alumnum,*
> *viribus ingenii his, corporis ille, valet.*

This can be translated as:

> *From Gothia were recently sent the mighty limbs of a giant,*
> *But in spite of their size, they lack the power of reason.*
> *Another plant grown in this fertile soil has now come to Upsala;*
> *While the former had bodily strength, he has that of the mind.*

Some years later, Swedenborg had occasion to reexamine the giant's
bones. From findings of fossil sea animals on the Swedish mainland, he had
drawn the conclusion that the water level must once have been much higher
than in his day. Swedenborg, then in his early thirties, had yet to turn to the
mysticism for which he became famous. He soon realized that the bones of
the Swedish giant had once belonged to a whale. He was perfectly correct,
and the still existing bones of the "Swedenborg whale," as it has been called,
have also interested modern cetacean biologists, since they are probably
from a subspecies now extinct, resembling, but not identical to, the Greenland whale.

The Decline of Gigantology

After the publication of Sir Hans Sloane's paper, traditional gigantology
went into a steady decline during the rest of the eighteenth century. But its
supporters did not give up easily. The Jesuits had always been enthusiastic
gigantologists, and as late as 1764, the Spaniard Father Tarrubia wrote a
long treatise titled *Giganthologia*, gathering many fabulous old stories as evidence of the literal truth of the Old Testament. It is more surprising that
when, five years later, the French naturalist Claude Nicolas Le Cat read a
memoir on giants before the Academy of Sciences of Rouen, he too put more
value on the authority of the old gigantologists than on that of Sir Hans

Sloane, of whose work he was well aware. Furthermore, several of Sloane's own countrymen were apparently quite unconvinced by his arguments. Several old-fashioned eighteenth-century antiquaries and county historians repeated old tales of giants' skeletons found underground, using them as evidence for the great antiquity of the part of England they were writing about.

In Scandinavia, gigantology also died hard. As late as 1763, a thesis on gigantic remains was published at the University of Upsala by the student Wilhelm Gustaf Zetterberg. Since this credulous youth, who must have been extremely badly advised by his supervisor, accepted even the remains of King Teutobochus as genuine giant's bones, he was harshly criticized in the newspapers. This contretemps did not prevent Archdeacon Tiburtz Tiburtius from sending another treatise on giants' bones to the Swedish Academy of Sciences, in whose *Transactions* it was published in 1765. A distinguished anatomist, Professor Roland Martin, who had himself seen spurious giants' bones in several churches, objected to this revival of an old fallacy, but the archdeacon persevered. He even sent parts of the skeleton to Stockholm as tangible proofs for the academians to examine. The Lutheran archdeacon also proved a worthy follower in the tradition of far-fetched theological speculations on this subject. His opinion was the direct opposite of that of the French academian M. Henrion: the Scandinavians had once been mighty giants, and their decline in stature was the fault of the Catholic Church. The christianizing of the Nordic heathens had brought an end to the giants' restful, sedentary lives and their diet of horse flesh and other kinds of strong, nourishing food. Instead, they were forced into hard labor, extensive fasting, and unhealthy, foreign food. To clench the argument, the archdeacon quoted the works of Olaus Rudbeckius, who had once seen a peasant boy nearly eight feet tall. When he was asked how he could have grown so tall, the boy answered that he had been breast-fed until he was seven years old and that he never got out of bed before noon; furthermore, he had not done any work before the age of eighteen.

One of the last supporters of traditional gigantology in America was the Reverend Ezra Stiles, president of Yale and one of the foremost American scholars of his time. He was the grandson of Edward Taylor, who had written the poem on the Giant of Claverack, and he had inherited his papers, several manuscripts on giants among them. Apparently, there was a legend among the Indians living near Claverack that a huge giant had once lived in these parts. He would chase the bears until they climbed up into high trees and then shake them down like ripe plums and devour them. When he felt like eating fish instead of bear's grease, the giant waded into a deep river and speared some sturgeons to roast on his fire. Ezra Stiles eagerly collected reports of other gigantic remains being found in America, receiving a good many bones, teeth, and manuscript descriptions from enthusiastic clerical colleagues all over the country. He had a passion for old-fashioned gigantology, which puzzled those of his scholarly contemporaries who kept abreast of the debate on this subject in Europe. In 1781 General Washington and his staff had examined a large find of huge bones and teeth near New Windsor, New York. Although it is recorded that the general presumed that the bones were from some large quadruped, Ezra Stiles stubbornly maintained that they were of human origin. Three years later, Thomas Jefferson wrote to Stiles to inquire why he disagreed with Buffon and other leading European naturalists in the question of the existence of giants. Ezra Stiles was adamant; he listed many findings of huge bones in America and Europe. Since such large animals no longer existed in his day, he presumed them to belong to an extinct race of bipeds of an immense stature.

At the turn of the nineteenth century, the rise of scientific paleontology had rendered the old-fashioned gigantology obsolete. The pioneer research of men such as John Hunter and Georges Cuvier gradually undermined the church's influence over this branch of science, and broke the ground for the early evolutionists. In his impressive *Des ossemens fossiles*, Baron Cuvier had occasion to reexamine several of the old giant's skeletons that were still kept in churches and private collections. Whereas Cotton Mather's American giant was a mastodon, Dr. Scheuchzer's antediluvian man proved to be a remarkably well preserved fossil giant salamander. Even in Baron Cuvier's time, there were objections from religious traditionalists, who maintained that the fauna in a certain territory had been unchanged since Creation and that it was impossible for any animal species to become extinct. If there were no elephants in Europe at present, then these bones must not be those of elephants. In the mid-1800s, this obstruction was also removed, and Darwin's breakthrough made it clear both that animal species could become extinct and that the species now living were related to those that had died out. The findings of fossil whales and pachyderms on the European mainland could now be explained, and traditional gigantology lost its last support. But the last of the giants in the earth had not yet been discovered.

The Cardiff Giant

On October 16, 1869, two workmen were digging a well at "Stub" Newell's farm in Cardiff, New York, when one of them hit a rock. They tried to dig it out of the ground and got the surprise of their lives when they saw that it was a ten-and-a-half-foot-long petrified giant. The newspapers were soon full of this amazing discovery, and thousands of people came to see the giant. Soon, Newell and his partner, the cigar manufacturer George Hull, put a tent over it and charged fifty cents for admission. They were successful beyond their wildest imagination: on one Sunday alone, not less than twenty-six hundred tickets were sold. The majority of the giant's admirers were gullible curiosity seekers who imagined that the question of antediluvian giants was now definitely solved, but Dr. Amos Westcott of Syracuse and several other medical men and academics also accepted it as a petrified man. An imaginative scholar, Professor Alexander McWhorter of Harvard University, instead believed it to be the statue of some heathen idol. He thought he could decipher some Phoenician characters on one of its arms and postulated that this people had explored America long before Columbus. It was reckoned that more than fifty thousand Americans had seen the Cardiff Giant's *lit de parade*; one of them was Ralph Waldo Emerson, who was apparently much impressed by it.

After a while, Newell and Hull thought it prudent to sell a three-quarter

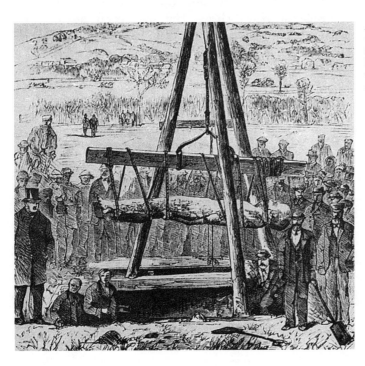

Fig. 4. The Cardiff Giant is hauled up from its grave. From a late nineteenth-century pamphlet in the author's collection.

interest in the giant, and a consortium of wealthy Syracuse citizens, the aforementioned Dr. Westcott among them, purchased it for thirty thousand dollars. The Cardiff Giant was then taken on tour, from Syracuse to Albany and New York. The legendary Prince of Humbugs, P. T. Barnum, visited the exhibition in Albany and offered to purchase the giant. The tough old "Stub" scorned his offer, but Barnum was too clever for him. The wily showman had an artist take a wax impression of the giant, from which a full-scale plaster model was made; when the Cardiff Giant arrived at New York, Barnum put his model on display as well, posting a barker outside to yell that Barnum had the real giant and that the other one was a fake. When the outraged "Stub" applied for an injunction to stop Barnum's deceit, the judge declared that he would issue one as soon as they were able to prove that their own giant was a genuine petrified man. Unfortunately, proof was getting increasingly difficult, since several scientists of repute had declared it a humbug, among them the paleontologist Dr. O. C. Marsh of Yale University. In February 1870 the bubble definitely burst. Two sculptors admitted that George Hull, the disreputable Binghamton cigar manufacturer, had hired them to make the giant from a huge gypsum block. Hull had got the idea after a quarrel with a methodist revivalist preacher named Turk about the existence of antediluvian giants. George Hull was greatly amazed at the preacher's naivety, and his original plan was to make a giant and bury it as a practical joke, to see the preacher's reaction when it was found. He was a thorough, clever man, and the task of making a lifelike giant challenged him. The sculptors worked for several months, using Hull himself as a model.

Fig. 5. The Cardiff Giant is exhibited. An illustration from a late nineteenth-century pamphlet in the author's collection.

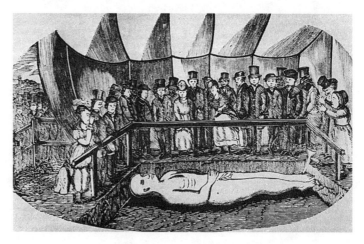

Hull later spent many hours finishing his work, sandpapering its surface and hammering it with a block of wood full of darning needles, to simulate the pores of the skin. The giant's darkish hue was due to the application of a gallon of sulphuric acid. The humbug makers later buried the giant at Newell's farm, and some days later, "Stub" ordered his workmen to dig a well at the same site. The considerable economic gain was a pleasant surprise for the two hoaxers; another triumph was that the Reverend Mr. Turk and his colleagues were completely fooled. They paid money to be allowed to hold prayer meetings in the tent before their huge idol.

The Cardiff Giant humbug reflects the sorry state of knowledge about natural history among nineteenth-century Americans. The hoax has not a few parallels with the King Teutobochus scandal in France, except that the Cardiff Giant was never taken to see the president, and the facetious pamphlets written about the American colossus cannot be compared to the rapier wit of the belligerent French pamphleteers. Like King Teutobochus, the Cardiff Giant has been kept for posterity; after a number of owners, it was purchased by the Farmer's Museum in Cooperstown, New York, where it can still be seen today.

❧ Legends of giants can be found in the lore of peoples all over the world. They usually depict the giants as stupid savages, dressed in hides and armed with clubs. Some ethnologists have suggested that the "giants" were mythologized barbarian tribes, existing on the outskirts of early civilized cultures; people were likely to exaggerate the size of formidable savage opponents met in battles and skirmishes. Many of the ancient fables about gigantic men may well have originated in such a way.

The church played a role in perpetuating the myths. From the vague and sometimes contradictory statements about giants in the Old Testament, an elaborate system of theology was constructed, giving the evil giants quite an important part in biblical history. The odd notions of Annius of Viterbo and Jean Lemaire brought this fairy-tale mythology well into modern times, linking the biblical giants with local traditions and the finding of huge fossil bones and teeth. The church had nailed its colors to the mast on the gigantology question, and then van Gorp's *Gigantomachie* dealt the doctrine an unexpected blow, for it was particularly well researched and its arguments were conclusive. To hold the tide against van Gorp's heretical opinions, theological countertheses were published, and the churches were kept well stocked with giants' bones. Several famous European cathedrals and churches, among them the Stephansdom in Vienna, Saint Peter's in Leipzig, and the Riddarholm Church in Stockholm, once housed the fossil bones of pachyderms and whales, and these huge relics were revered by millions of people. Sir Thomas Molyneux commented in 1699 that during the last century, the accumulation of gigantic remains was pursued with such diligence that "I confess there is hardly a considerable collection of [natural curiosities], or a printed description of a *Musæum* extant, where some part or other of a Giant is not to be met with."

The first conclusive proof that the huge skeletal remains were not those of gigantic human beings came from Sir Hans Sloane. Although Johan van Gorp had drawn similar conclusions 157 years earlier, he had based them on the examination of a single spurious giant's tooth. Furthermore, the authority of the church was against him, and the contemporary naturalists were highly conservative and bound by authority. A great hindrance to explaining the findings of fossil pachyderms and cetaceans on the European mainland was the dogma, then universally accepted, that no animal species could become extinct and that the fauna within a territory had been unchanged since Creation. Johan van Gorp was unable to free himself from these limitations; he could state merely that the giant's tooth of Antwerp had belonged to an elephant; he could not explain how this elephant had come to Belgium, except to propose that it might have been a combat elephant taken there by the Romans. Sir Hans Sloane was also puzzled by the frequent findings of fossil pachyderms in Europe. He quoted the opinion of Johan van Gorp and Count Marsili, that the elephants had been brought there by the Romans, but from his own observations, he could not accept it. The fossil bones were confined to deeper strata in the ground, implying that they were much older, and furthermore, the erudite Sir Hans knew that the Romans used to remove the tusks from their dead elephants, whereas the skeletons usually had their tusks entire. His own theory was that the skeletons had been brought there by the force of a universal deluge.

It was not until men such as Blumenbach, John Hunter, and Baron Cuvier had disproved the old dogma of the constant distribution of animal species that a theoretical basis was provided for finally discarding traditional gigantology. Although the church could not stop these new ideas from being universally accepted among men of science, it certainly delayed their diffusion into the body of general knowledge, as the burlesque tale of the Cardiff Giant testifies. Several European churches sported gigantic remains well into the nineteenth century. Even today, the giants in the earth have not been laid entirely to rest. Some enthusiasts have reinvoked the existence of huge skeletal remains as evidence that the earth was once populated by a race of mighty giants, who arrived in huge UFOs. Furthermore, some fundamentalist religious groups, taking the biblical story of creation literally, have recruited the giants in the earth for use in their books and pamphlets. As late as 1986, strict creationists interpreted a cluster of fossilized tracks near Glen Rose, Texas, as those of a dinosaur and a gigantic human being. They took this evidence as proof that the earth was created only six thousand years ago, complete with all animal species, and that before the Deluge, biblical giants and dinosaurs had walked the earth together. Several American paleontologists of repute objected to this Fred Flintstone version of prehistory gaining popular credence, and they conclusively demonstrated that all the Texas tracks were of saurian origin. Otherwise, this startling theory might have provided a novel explanation of the extinction of dinosaurs. Surely, they could not have survived long among the giants, being continuously stepped on and crushed by mistake, roasted under the sun as snacks, or kept in small cages like parakeets in the weird Jurassic Park of King Og and his gigantic companions.

Apparent Death and Premature Burial

    There was a young man at Nunhead,
Who awoke in a coffin of lead:
"It is cosy enough,"
He remarked in a huff,
"But I wasn't aware I was dead."

The Perils of Premature Burial (1905)

T HE POPULAR LITERATURE OF EARLY NINE-
teenth-century Britain was steeped in blood and gore. The cheap
horror novels, also known as "bloods" or "penny dreadfuls," written
by the imitators of Mrs. Radcliffe and "Monk" Lewis, were everywhere, and
the popular magazines had titles such as the *Terrific Register*; the *Record of
Crimes, Judgments, Providences, and Calamities*; and *The Ghost*. Their articles
dealt with murder, torture, spontaneous human combustion, incest, molten
lead poured into wounds, and the life stories of Sawney Beane, the Scottish
cannibal mass murderer, and Sweeney Todd, the Demon Barber of Fleet
Street. Another pet subject was the horrors of premature burial, described
in hideous detail.

    One of these gruesome articles, headlined "Terrible Event!," described
the sad fate of Baron Hornstein, a popular courtier in Bavaria, who had been

buried with great ceremony after his sudden demise, from apoplexy, at an early age.

> Two days after the Funeral, the workmen entered the Mauso-leum, when they witnessed an object which petrified them with HORROR!!! At the door of the sepulchre lay a body covered with *Blood*! It was the mortal remains of the *favourite of Courts and Princes*! The Baron had been BURIED ALIVE!!! On recover-ing from his Trance, he had forced the lock of his Coffin, and endeavoured to escape from the Charnel-house! It was impos-sible! He therefore, in a fit of desperation, had dashed his brains out against the wall!!! The Royal Family, and indeed the whole city, are plunged in grief at the horrid catastrophe.

Even a regular reader of the most dreadful pseudogothic periodicals must have felt a *frisson* of horror. The popular fear of premature burial was at its greatest in the early Victorian era, at times virtually amounting to mass hysteria. Those who wanted to ensure themselves against being buried in a state of apparent death could order their coffins equipped with the Bateson Life Revival Device, an iron bell mounted in a miniature campanile on the lid of the casket, the bell rope being connected to the presumed corpse's hands through a hole in the coffin lid. Bateson's Belfry, as it was called, was of course most useful before burial, but those wealthy and cautious enough could have the apparatus buried with them. The Victorian gentleman may have hoped that, should he revive in his padded coffin, the sinister under-ground knelling from the belfry would call some attentive verger or grave-digger to the rescue. The coffin might then be unearthed and the creaking lid opened by a polite servant, with the words, "You rang, Sir?"

෴ Before the 1740s people all over the world seem to have given but little thought to the gruesome possibility of being buried alive. Death was consid-ered a supernatural, irrevocable event; either one was alive or one was dead, and the crossing of the border was instantaneous. It is true that already in antiquity, there was some awareness that the diagnosis of death was not al-ways straightforward. According to Apuleius, one of the legendary feats of Asclepiades was to discover that a man laid out as a corpse was not really dead. Pliny in *De funeribus romanorum* states that the consul Acilius Aviola and the praetor Lucius Lamia both awoke on their flaming funeral pyres and that the attendants could save neither of them from a most horrible death.

According to a gravestone in Milan, François de Civille was declared dead in 1562 and buried in the graveyard of that city. Six hours later, his brother sensed in some strange way that something was amiss, and François was disinterred and revived. He lived for seventy more years, finally dying at the age of 105, from a chill contracted while "serenading the lady of his heart all night long." Some sixteenth- and seventeenth-century physicians were aware of the danger of hasty funerals. Already the old Salmuth recommended caution in these matters; in particular, the bodies of women known to be of a nervous and hysterical disposition were to be left above ground for three days before burial. In 1670 the Germans Theodorus Kirchmayer and Christophorus Nottnagel pointed out the difficulty in distinguishing real from apparent death and maintained that it was wise to delay the funeral for some days when there was doubt. Dr. Lancisi, first physician to Pope Clement IX, reported in a book on various aspects of sudden death that he had once seen a presumed corpse sit up in his coffin during the funeral mass in a church in Rome. This experience naturally made him doubt the current principles for declaring people dead. Neither of these works was widely read, and neither did more than hint at the horrors of premature burial.

The thesis on the uncertainty of the signs of death written in 1740 by Jacques-Bénigne Winslow, professor of anatomy at the Jardin du Roi, also, at first, seemed destined to oblivion on the shelves of forgotten academic treatises. Winslow was a distinguished anatomist, a native of Denmark and a relative of the celebrated Danish anatomist Nicolaus Steno; he had emigrated to France in the late 1600s. He claimed that he had twice, once as a child and once as a young adult, narrowly escaped being buried alive. The disturbing conclusion of this thesis was that the current principles for declaring people dead were highly unreliable and that live people were frequently buried by mistake.

The Paris physician Jean-Jacques Bruhier d'Ablaincourt was one of those who read Winslow's Latin thesis. He felt that these dramatic and far-reaching conclusions should become public knowledge, and he decided to translate it into French, with considerable additions of his own. Public interest was instantaneous, and in France, Bruhier's *Dissertation sur l'incertitude des signes de la mort* went into three editions. This medical best seller was translated into German, Italian, Dutch, Spanish, Danish, and Swedish, and its triumphal tour throughout Europe left a wake of horror and dread.

In 1746 an English translation of Bruhier's book was published, titled *The Uncertainty of the Signs of Death* and lacking the author's name on the title page. It may well have been an unauthorized translation, meant to cheat the French authors of the profits from this best seller, which went into at least three English editions, some of them printed in Dublin. The book promised to present "a great Variety of amusing and well-attested Instances of Persons who have return'd to Life in their Coffins, in their Graves, or under the Hands of the Surgeons." Winslow had heard of a rather "amusing" instance from the prior Joseph Mareschal: this cleric had once seen a woman sit up in her (fortunately open) coffin, when the pallbearers passed through rue Jean Robert on the way to the graveyard. Winslow and Bruhier did not consider the absence of respiratory movement and arterial pulse to be infallible signs of death. Only putrefaction and the appearance of "livid spots" were sure signs that the individual had really expired. They maintained that no lifeless patient who could not be safely diagnosed as dead was to be shrouded and put into a coffin. Instead, the presumed corpse should remain in a warm bed, and vigorous attempts should be made to resuscitate it. The individual's nostrils were to be irritated by introducing "sternutaries, errhines, juices of onions, garlic and horseradish." Some advocated tickling the nose with the quill of a pen; others preferred thrusting a sharp, pointed pencil up the corpse's nose. The gums were to be rubbed with garlic, and the skin stimulated by the liberal application of "whips and nettles." The intestines could be irritated by the most acrid clysters, the limbs agitated through violent pulling, and the ears shocked "by hideous Shrieks and excessive Noises."

The corpse that withstood these brutal methods of resuscitation had even more unpleasant experiences ahead. The soles of the feet were cut with razors and long needles thrust under the toenails. Lancisi recommended that a red-hot iron be applied to the soles of the feet; Winslow would have boiling Spanish wax poured on the corpse's forehead; a drastic French clergyman advocated that a red-hot poker be thrust up the unfortunate corpse's rear quarters as a last resort.

Although these violent and barbaric measures seem ludicrous today, more worthy of the torture chambers of the marquis de Sade than the mortuary of a French hospital, Jacques-Bénigne Winslow was guided solely by humanitarian feelings. Bruhier's additions to the book were rather more sensational. He presented several cases of fatal or near-fatal mistakes in detecting death. One of them concerned Jean de Lavaur, a nobleman of Neufchâtel, who was brought back to life in his coffin when his physician blew powdered pepper into his nostrils. Bruhier was apparently of the opinion that a person might live for quite a long time in the absence of both heartbeat and respiration. He mentioned the example of a Swedish gardener from Drottningholm Castle outside Stockholm, who had fallen through the ice and become lodged underneath. He survived for sixteen hours before a boatman thrust a hook into his head with great force, to pull him up alive. The submarine gardener was received in audience by the queen of Sweden, who examined the mark of the hook in his head and gave him a pension for life. According to other idle old tales, a German lady had lived three days under the waterline, and the nephew of the archbishop of Cologne had taken a fifteen-day dive before emerging alive and well; the record was held by a Swiss youth, who had spent seven weeks under water without any ill effects. After reviewing these and other astounding examples of the extreme capacity of the human body to survive without oxygen, Bruhier was unwilling to accept any sign of death except putrefaction. His book was extremely widely read, and all around Europe, many people put clauses in their wills to warn their elderly and shortsighted family practitioners to take care and prevent their greedy heirs from having them put into the earth too hastily. Some even demanded that vigorous attempts be made to revive them before their presumed corpses were buried, according to the schemes of Winslow and Bruhier.

Several French and German medical men objected to Bruhier's ideas, none more vigorously than the Paris physician Antoine Louis. He held Bruhier responsible for the premature burial hysteria in France. His ill-researched book had spread "des histoires hazardées ou ingénieusement controuvées pour amuser les femmes et les enfans" — "silly stories cleverly concocted to amuse women and children." In particular, he objected to the use of putrefaction as the sole sign of death, since delaying burial meant that putrefying bodies would remain in people's houses for a considerable time. Surely the emotional distress to the surviving relatives was not justifiable. Furthermore, the "very fetid odor" of the corpse might poison the living people in the household.

At this time, the French countryside was full of quacks, wise women and itinerant charlatans, who were serious rivals of the professional medical men. The French peasants were, not unreasonably, of the opinion that if the doctor could not tell a living person from a dead one, he was good for absolutely nothing. Bruhier's ideas about the uncertainty of the signs of death thus threatened the authority of the medical profession at large. M. Louis believed that rigor mortis and certain changes in the eyes were even more certain signs of death than putrefaction, which might occur in a festering, gangrenous limb. He himself used a remarkable apparatus, made especially for the purpose of awakening those who were apparently dead. One of the pipes of this curious contraption, which rather resembled an oversized set of bag-pipes, was to be inserted into the anus, and another was connected, by way of a powerful bellows, to a large furnace full of tobacco. Another of these machines was used in London, and Louis prided himself that several Dutch and German physicians, anxious to know the latest advances in clinical medicine, had been to France to see these enemas of tobacco smoke being administered to the hapless corpses at the Paris morgue.

In 1745 Bruhier was received in audience by King Louis XV. His Majesty was apparently much impressed by the Paris physician's gruesome lecture, since he made a half promise that the state would establish and finance a corps of mortuary attendants whose job it would be to save the seeming dead. When the king's ministers had considered the costs of this reform, they persuaded Louis XV to postpone it indefinitely, to Bruhier's chagrin. In spite of this disappointment, Bruhier and Winslow were the winners of the French debate on the uncertainty of the signs of death; the majority of medical men took their side in the controversy, disregarding the objections of Louis and others. For example, the article on death in the famous *Encyclopédie* of Diderot and d'Alembert closely adhered to the teachings of Bruhier. The most important and beneficial effect of Bruhier's work was that many European countries instituted regulations that delayed burial for one or more days after death. At Jean-Jacques Bruhier's death in 1756, a friend of his wrote an admiring epitaph, containing the lines:

> *Bruher, ton immortel ouvrage*
> *Ouvre les yeux à bien des gens*
> *Sur l'abus, le cruel usage*
> *D'enterrer les morts tout vivants.*

By the late 1700s premature burial had become one of the most feared dangers of everyday life, and a torrent of pamphlets and academic theses were dedicated to this subject by writers all over Europe. The influential German physician Cristoph Wilhelm Hufeland agreed with Bruhier that putrefaction was the only certain sign of death. He also suggested that a particular state of deep unconsciousness, the *Scheintod*, or death trance, always preceded real death. The *Scheintod* was indistinguishable from real death, and it could last for quite a long time. Although the apparently dead person seemed to lack arterial pulsations, muscular reflexes, and respiratory movements, the death trance was not always fatal. Vigorous resuscitation could save the individual's life.

It seems that these trancelike states really existed at the time and that they occurred fairly frequently, usually in younger women. In most cases their perception was unimpaired, enabling them to hear and understand what was happening, although they were powerless to move or cry out, even when they were put in a coffin and prepared for burial. Sometimes quite trivial matters aroused them from the trance. One lady woke up after hearing the voice of a childhood friend, another after a lutist had played one of her favorite melodies. M. Chevalier, a Paris surgeon who was known as a great piquet player, was aroused from a deep trance, diagnosed by some as death, after one of his friends called out the piquet commands "Quint, quarante, point!" These remarkable nineteenth-century cases were of course widely publicized at the time. Some of them, but by no means all, can be explained as unusually persistent states of hysterical conversion.

In the late 1700s and early 1800s, there was no shortage of books, pamphlets, and articles on apparent death and premature burial. Many people were frightened half out of their wits, and they left legacies to their family physicians to safeguard themselves against this gruesome fate. The writer Harriet Martineau left her doctor ten guineas to see that her head was amputated, and the antiquary Francis Douce left two hundred guineas to the surgeon Sir Anthony Carlisle to see to it that his heart was taken out after death. Miss Beswick, an old lady living in Manchester in the late eighteenth century, left twenty thousand guineas to her family physician, Dr. Charles White, on the condition that she was *never* to be buried; her body was instead to be embalmed and kept in the doctor's collection of anatomical preparations. Every day for several years the doctor and two reliable witnesses were to lift the veil and survey the countenance of this macabre resident patient for any signs of life. Later, the doctor put Miss Beswick's mummy inside an old clock case, opening the clock once a year to see how his favorite patient was doing. Her eccentric will made the immortal Miss Beswick quite a celebrity. Thomas de Quincy and a highborn lady friend once went to Manchester to see Miss Beswick, but Dr. White refused to show them either the mummy or the clock, although he politely showed them the other scientific curiosities in his house in Lower King Street. To the end, the doctor loyally fulfilled his side of the deal, but after his death, Miss Beswick's mummy was taken to the Manchester Museum, where it was publicly exhibited in the entrance hall. It was finally decently interred in 1868.

The warnings of Bruhier and Hufeland were taken very seriously in the German states. Many cities employed inspectors of dead bodies to examine and, if doubt arose, attempt to resuscitate every corpse within their area of jurisdiction. Several German cities built *Leichenhäuser*, or waiting mortuaries, where the corpses were to be kept until they showed signs of putrefaction. These establishments were staffed by a matron and several porters and nurses; one of the town physicians was kept "on call" should any of the patients show signs of life. Every corpse was placed on a comfortable stretcher or bed, with strings tied to their fingers and connected to a large bell, whose tolling could awaken, if not the dead, at least the sleepy night watchmen of the waiting mortuary. Dr. Hufeland personally supervised the erection of a waiting mortuary in Weimar, a copy of a French establishment of the same kind. In Frankfurt am Main, a wealthy local philanthropist with a horror of being prematurely buried donated money to build a particularly luxurious *Leichenhaus*. There were two wards with twenty-three beds in each, under the continuous watch of nurses and porters. The doctor made his rounds every day. White gloves, placed on the hands of every lifeless patient in this ghoulish hospital, were connected to a powerful gong through an intricate system of strings. Other waiting mortuaries were built in Berlin, Weimar, Augsburg, and Mainz. There were ten of them in Munich, catering to various religious persuasions.

In 1792 Mme Necker, who herself had written a pamphlet on premature burial, submitted a proposal to the National Assembly recommending that a waiting mortuary be built immediately in Paris. The literary lady was quite unable to convince Messieurs Robespierre, Marat, and Danton of the urgency of this reform, however, and nothing came of it. In 1801 there was another plan to build six *temples funéraires* in Paris, but Napoleon Bonaparte, who apparently did not share the popular fear, turned it down. Throughout

Fig. 1. The interior of a Munich *Leichenhaus* in the 1890s, from the 1905 edition of Tebb and Vollum's *Premature Burial*. Reproduced by permission of the British Library, London.

the early 1800s, many French philanthropists advocated the building of *dé-pôts mortuaires*, and a few were built, but the idea never seems to have caught on in a big way beyond the borders of Germany. Plans to build waiting mortuaries in Copenhagen and Amsterdam were never acted upon. One was built in The Hague in the 1830s, and it still stands, but although this *Schijn-dodenhuis* was fully equipped with bells and strings, it seems to have been used as a regular mortuary most of the time. Not infrequently, the corpses in the German mortuary hospitals did indeed ring their bells, but their movement was the effect of decomposition. There is no case on record of any of the denizens of these institutions ever coming to life.

In the early 1800s research on the clinical and physiological signs of death had made but little progress since the time of Bruhier and Louis. To encourage research, several philanthropic French and German magnates contributed large sums of money to prize contests dedicated to solving the problem of distinguishing real from apparent death. There was certainly no shortage of contestants or of crackpot ideas about how to tell corpses from living people. Strong smelling salts and pinches of foul-smelling snuff should be put into the corpse's nostrils, fanfares blown on a powerful bugle next to the unprotected ear, and the body's most sensitive areas rubbed with stiff, prickly brushes. A Frenchman invented a pair of strong pincers with which to pinch the presumed corpse's nipples, and an Englishman named Barnett recommended that the arm of the corpse be scalded with boiling water; if a blister appeared, the individual was still alive. According to some fanatics, these macabre activities were to be kept up for days or even weeks before the poor corpse could be left in peace in its coffin. Another odd notion was that a long needle, with a flag attached to one end, should be thrust into the heart of the apparently dead individual; the flag would wave merrily if the heart was still beating! A Swedish writer recommended that "a crawling insect" be conveyed into the corpse's ear. Another eccentric inventor believed that the relaxation of the intestinal sphincters after death might prove useful as a sign of death. He presented an apparatus rather like that used by Louis in the 1750s, but with the exactly reverse application: a nozzle was forced down the corpse's throat, and a powerful air pump applied by two strong men; the grotesque scene that must have followed can be imagined by those familiar with dead bodies, but remains mercifully hidden to the laymen.

The ideas of these cranks did not find favor with the prize boards, and it was not until 1846 that any major sum of money was paid out. A French doctor named Bouchut received fifteen hundred gold francs for his suggestion that the newly invented stethoscope be used to detect the cessation of heartbeats. This reward went to the right man, since Bouchut was of the firm opinion that the stethoscope could make the lack of heartbeat certain, leaving no doubt and allowing the dead body to be buried without further delay. He condemned the waiting mortuaries as needless and macabre and considered his countrymen's great fear of being prematurely buried to be greatly exaggerated, fueled by the many inaccurate and sensation-seeking pamphlets on this subject. In one of these, a certain M. H. Le Guern claimed that more than fifteen hundred people were prematurely buried each year. One French lady writer who had strong feelings on the subject recommended that all corpses be kept above ground for at least eight days, after which their arms, legs, and heads were to be chopped off before burial. Although Bouchut succeeded in convincing many of his professional colleagues throughout Europe, he had little success among the French public. They preferred the sensational newspaper accounts to his sober reasoning, and in the 1860s the fear of premature burial rose to new heights. The Chamber of Deputies was frequently appealed to, and grandiose schemes to build waiting mortuaries devised. Foreigners, particularly Englishmen, were afraid of traveling alone in France during this time, fearing that they would awaken under ground after a mere fainting fit.

In the early 1800s books of anecdotes about apparently dead people coming to life were eagerly read by all and sundry. The apparently dead person returning from the tomb was a hero who had escaped the clutches of the Grim Reaper; in particular, the prematurely buried woman — the sleeping beauty who could not be awakened from her trance — was a subject of gruesome fascination. Several popular tales on this theme were already current in the 1700s. They occurred in many variations and can be presumed to be purely invented. Nonetheless, they were important, for these idle tales were uncritically repeated by the majority of later writers on premature burial, from the somber Latin pages of the doctoral thesis to those of the vulgar penny dreadful.

The oldest, and most famous, of these popular legends, was that of the lady and the ring, which occurred in at least seven versions, with different names for the protagonists and varying geographical locations in France, Italy, and Germany. The main theme is that the wife of a wealthy gentleman suddenly dies; she is buried in her best clothes, with a large gold ring on her finger. In the night a servant digs up the coffin to steal the ring, but he is unable to get it off her finger. He then tries to cut the finger off with a dagger, but at the first stroke of the knife, the recumbent lady awakens with a piercing

Fig. 2. The apparently dead lady awakens when the robber tries to cut off her finger to steal a ring. A drawing from *The Uncertainty of the Signs of Death*, the pirated English version of Bruhier's book, published in Dublin in 1746. Reproduced by permission of the British Library, London.

scream. The lady is thus saved and reunited with her grieving husband. The fate of the grave-robbing servant varies in the different versions of the tale: in the more benevolent interpretation, he is given a reward for having saved the life of his mistress, and in the vindictive one, the thieving menial is struck dead by terror when the presumed corpse awakens from her trance.

A second, no less sentimental eighteenth-century anecdote tells the story of two young lovers who are forbidden to marry by her stern father. He prefers a wicked nobleman (or tax collector) as a son-in-law, and after she is married to him, the young lady falls into a decline, as befits the heroine of a French melodrama. Finally, she dies from the effects of a broken heart. Her grief-stricken ex-fiancé visits her tomb in the family vault, planning to cut his throat with a razor there, but the girl awakens from her deathlike trance just in time to stop him. They visit the wicked nobleman, and he drops dead on seeing the presumed ghost of his wife. The French tax collector, who is the villain in the other version, is made of sterner stuff. Not only does he survive the shock of his wife's return from the dead, but he actually tries to reclaim her as his lawful wife, thereby forcing the young couple to flee to England.

A third, no less unlikely tale is that of the lecherous monk. A young French gentleman is forced to become a monk by his religious parents, though he has no vocation. On his way to the monastery, he stops at an inn. The innkeeper persuades him to watch over the corpse of his beautiful young daughter, who died the day before. At the sight of her, the monk "forgot the sanctity of his vows and took liberties with the corpse." After the monk's departure, the apparently dead girl comes back to life. As fate would have it, the monk returns to the inn nine months later, and to his great surprise, sees the girl alive, and with a newborn child in her arms. He at once tells the parents that he is the child's father, casts off his monkish gown, and offers to marry their daughter. The innkeeper and his wife are delighted to have this handsome, well-mannered young man as their son-in-law, even though the lecherous ex-monk has confessed, in so many words, to having raped their daughter while he presumed her to be a corpse.

The British public did not have to rely solely on the importation from Germany and France of these silly stories about prematurely buried people being saved from the tomb. In 1816, *The Danger of Premature Interment*, by Joseph Taylor of Newington Butts, appeared in London. This imaginative author had previously written on apparitions and ghosts, "canine gratitude," and "anecdotes of remarkable insects." His diligence in seeking odd reports of apparently dead people coming to life is admirable, but his conclusions from them do not reflect creditably on his medical knowledge. The Lady and the Ring tale was of course reproduced in his collection, together with a rather amusing English variation on the same theme, said to have involved Sir Hugh Ackland of Devonshire. This worthy had died from a fever, as it seemed, and two footmen sat up with the corpse, which had been laid out in a coffin. One of them remembered that his late master "dearly loved brandy when he was alive," and he poured a glass of brandy down the corpse's throat. Sir Hugh immediately responded to treatment and came spluttering to life. He survived for several years, and the impudent footman was given a handsome annuity.

If Taylor's book was relatively balanced, John Snart's *Thesaurus of Horror; or, The Charnelhouse Explored*, which was published the next year, was the most ludicrous and gruesome example of this literary genre. The nature of its contents can be easily gleaned from the deplorable poem that introduced the book's main theme:

> *But if the fertilizing earth restore*
> *The dubious fragment of a borrow'd life,*
> *Can man's most desp'rate scuffle force the grave,*
> *Or must he, grappling, bathe himself in blood,*
> *And burst his eyeballs in the vain attempt!!!*

The British Library's copy of Snart's book has pasted into it a letter from Neariah Snart, the author's daughter, containing her solemn promise that her father was never to be buried alive.

John Snart also detailed the tragic fate of the Baron Hornstein, who dashed his brains out against the wall — a gesture much favored by desperate villains in French novels, though its actual performance would take the neck muscles of a prize bull and a skull the thickness of a porcelain teacup. Such a physique would have made the baron an interesting enough spectacle already during his life, but the *Almanach de Gotha* raises doubts about his very existence.

❧ While the fear of premature burial was abating in Germany and Scandinavia in the 1870s, it still reigned supreme in France. Furthermore, the old ghost of premature interment reappeared in both Britain and the United States, and several new books were published on both sides of the Atlantic. This development was partly related to the growth of the spiritualist movement, whose adherents worried that a spirit, which had left the body to lark about on its own, might have to return to it when it had been declared dead for lack of an occupant and put in a coffin deep in the ground. They gathered support for their ideas from the book *Premature Burial* by the Austrian Dr. Franz Hartmann, which was published in 1896. Hartmann combined the old errors about the death trance with the modern ones about spirits and astral bodies. Competent hygienists and physiologists derided his book, but the following year a worthier challenge to the medical establishment appeared. *Premature Burial* by the Englishman William Tebb and the American Colonel Edward P. Vollum was one of the most influential books on its subject; in more than four hundred pages it reviewed all aspects of the problem. The authors, no obvious cranks, conveyed an air of objectivity, and even medical men found the book hard to shrug off. Tebb and Vollum presented more case reports than ever before, ranging from the old idle tales of the Lady and the Ring and the Lecherous Monk, to the most gruesome stories, usually from newspaper sources, of apparently dead people eating their own hands, fingernails "encrusted" in the coffin lid, and prematurely buried women bearing children in their coffins. Some cases had more than a hint of humor. For example, an Indian missionary named Schwarz, lying in his coffin in the church, was wakened from his death trance when he heard his favorite hymn being sung outside; he made his resuscitation known by joining in the song!

In the late 1800s, many safety coffins were patented. One of the first and most successful was the aforementioned Bateson's Belfry, which appeared in the 1850s and made its inventor a wealthy man. Queen Victoria awarded George Bateson the Order of the British Empire for his services to the dead. According to some accounts, Bateson was later driven insane by his preoccupation with the horrors of premature burial. Not trusting even his own apparatus, he rewrote his will, asking to be cremated. Later, the wretched man began to doubt that his directions would be followed; in a fit of desperation, he doused himself with linseed oil and set himself ablaze in his workshop, preferring a premature cremation to the lingering risk of premature burial.

In the United States, not fewer than twenty-two patent applications were submitted for security coffins between 1868 and 1925. Their construction varied a good deal; one of the earliest, and probably most practical, models had a hollow tube and a rope ladder down to the coffin, which was equipped with a sliding door in its lid. The detachable tube could be removed, and the sliding door closed if, on later inspection, life was proven to be extinct. The majority of later designs relied on electrical signals that would trigger flags, bells, or rotating lights. The drawback was that mechanisms sensitive enough to yield to the most minute movement of the presumed corpse, could easily be set off by decomposition itself, leading to distressing and farcical false alarms. There were drawings for a coffin with an ejection seat, for use in vault burials, but it does not seem to have been manufactured.

Count Michel de Karnice-Karnicki, a chamberlain to Czar Nicholas of Russia, had once, he claimed, seen a young Belgian girl awaken in her coffin while the first shovelfuls of earth were being thrown upon it. Much affected by this experience, he decided to invent a mechanical apparatus to save the lives of others who were buried prematurely. After several years' work, he was able to present a particularly advanced security coffin, eponymously called Le Karnice, before an international audience at Sorbonne in 1897. The main idea was that every time someone was buried, a long tube was to be connected to the coffin, the end sticking up through the earth like the periscope of a submarine. The slightest movement of the "corpse" in its coffin would set into action an intricate mechanism, which would open the tube to admit air into the coffin. At the same time, a bell would ring, and a flag start waving on the top of the mechanism, which had some resemblance to an American mailbox. The noise and waving would, it was hoped, call the diligent vergers and gravediggers to the rescue. Apparently, Le Karnice worked quite well when tried out at the Sorbonne before a large gathering of foreign ambassadors, medical men, hygienists, and journalists. Perhaps it failed at some other trials in France, however, for the count's French agent, Emile Camis, expressed outrage that the philanthropic Russian had had to endure ribald comments in the French newspapers, and that the ignorant journalists had laughed at Le Karnice. It is to be hoped that the Count's intrepid assistant, who allowed himself to be buried alive, came to no harm! In spite of these difficulties, several medical hygienists of repute advocated the use of Le Karnice, among them Professor Richot of Paris. In 1899 the enterprising Camis took Le Karnice to the United States and exhibited it at 835 Broadway in New York. The American funeral directors were not particularly impressed by it. Probably word had spread from Europe that the mechanism was much too sensitive; it did not allow for even the slightest movement of the corpse during decomposition. The result was bizarre scenes at the graveyards, with bells ringing and little flags waving on the mailboxes. The grumbling vergers, who had to dig up the putrefying corpses, refused to have anything to do with Le Karnice and other security coffins like it.

After Tebb and Vollum had published their book, several doctors and hygienists, incensed at the revival of a popular medical fallacy, wrote books

or pamphlets in rebuttal. In Britain the dermatologist Dr. David Walsh had
some success with his book *Premature Burial: Fact or Fiction?* Although these
works may have convinced the members of the medical profession, the
American public preferred the sensational and horrific books on premature
burial, which remained best sellers well into the early 1900s. Between 1905
and 1908 an English society for the prevention of the burial of live people
edited a monthly journal titled the *Burial Reformer*. In the United States there
was a magazine called *Our Dumb Animals* that crusaded both for animals
threatened by cruel vivisectors and for apparently dead people facing the
horrors of a living inhumation; its distasteful stories were eagerly regurgi-
tated by several provincial newspapers. At length, in 1909 the *Journal of the
American Medical Association* reacted against this fueling of an old superstition,
and the editors were able to show that of three sensational examples of pre-

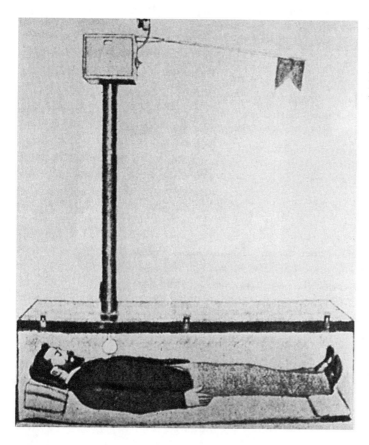

Fig. 3. A drawing of
Le Karnice from a
copy of Count de
Karnice-Karnicki's
undated American ex-
hibition pamphlet in
the author's collection.

mature burial recorded in *Our Dumb Animals*, two were complete fabrications and a third lifted from an article, in the same magazine, nine years previously.

After the First World War, the public fear of premature burial gradually diminished. Despite the occasional report of some presumed corpse that came to life in the morgue, or even in the coffin, the subject was losing its power of fascination. The trade in patent security coffins continued, rather surprisingly, although they were not given as much publicity as in the days of Count Karnice-Karnicki. According to an article in a French journal, a security coffin was still made in the late 1970s; although it cost £4,500 — as much as a motor car — several wealthy Americans purchased this super-coffin. It was luxuriously upholstered, provided a soft pillow for the head, and was large enough to permit the buried individual to sit up. The coffin's food locker was supplied with rations like those of the American astronauts. On the coffin's dashboard were controls for the oxygen, supplied by large gas tubes; the ventilation fan; the air pumps; the chemical toilet; the electric alarm bell; and the short-wave radio transmitter and receiver, for which an aerial stuck up above ground. Several extra gadgets, such as a small oven, a refrigerator, and a hi-fi cassette player, were optional.

❧ The traditional explanation of the great fear of premature burial is that Winslow and Bruhier started it all, but a more important question is what prompted them to do so. Some historians have postulated that the rising apprehension about premature burial in the early 1700s was related to the increasing use of coffins; these would give deceased individuals a private space in death, a close, impenetrable wooden shell, and anxious people would of course start wondering what might happen if the person in the coffin were not really dead. In fact, however, according to the historians of European burial customs, coffins had been in use since the fourteenth century and shrouds since the early Middle Ages. Already by the sixteenth and seventeenth centuries, burial without a coffin was reserved for paupers. It may have been that the use of shrouds and coffins became even more common in the early 1700s; furthermore, there is some evidence that by the late seventeenth century, people were putting precautions into their wills to safeguard themselves against premature burial.

It seems reasonable to relate the European fear of premature burial to the contemporary perception of death, which was changing from a purely mechanistic model to one involving a *process* of dying. The boundary between life and death was becoming unclear, and it was beginning to be accepted that a living person might have the appearance of death. The concept of death as a process is in fact correct, for unless death is very sudden — due, say, to decapitation, a bursting aneurysm, or a massive coronary embolism — the transition is gradual. In an aged individual dying from cancer, death creeps through the tissues, taking over one organ after another. The old concept of a death trance is not upheld by modern medical science, and no modern cases resembling those described with such care by early nineteenth-century physicians have been recorded.

As late as a hundred years ago, several writers gave alarming statistics on the number of people buried before they were dead. The Englishman J. C. Ouseley thought that twenty-seven hundred people were buried alive in England and Wales each year, and an equally imaginative French writer calculated that more than two thousand of his countrymen annually met a like fate. A Swedish physician rashly speculated that in his country, one out of ten people were buried before they were dead. The American attorney Marvin Dana, writing in 1897, claimed that an American was buried alive each week. A thorough review of the contemporary literature on apparent death and premature burial would rather support the opposing view of Bouchut, Walsh, and other skeptics that these figures were highly exaggerated. Many of the case reports reproduced by Hartmann and Tebb and Vollum are brief extracts from the newspapers, often told and retold by several papers before reaching the United States. Of the more detailed accounts, many are wholly unconvincing. Striving for sensational and horrific headlines, the writers made much of a cadaver found, for example, lying on its side; this position could just as well be consistent with the tilting of a coffin while it was being put into the grave. If the corpse's face was contorted or the arms and legs drawn up, an imaginative journalist might describe the horrid scuffle in the coffin, but neither of these changes is inconsistent with the natural decomposition of the body after death. A shattered coffin in a vault might, similarly, be due to the swelling of the corpse during putrefaction. Even the gruesome phenomenon of a dead pregnant woman bearing a child could be caused by the highly increased intraabdominal pressure during decomposition.

In several nineteenth-century reports, the prematurely dead individuals are said to have eaten their own fingers or even their entire arms; modern textbooks in forensic medicine have demonstrated that such bodies were probably attacked by rodents. In a much-publicized case from Kronstadt, the skeleton of a boy who had been buried fourteen years previously was found lying on the floor beside its coffin. The other coffins in the family vault were also empty and plundered. The premature burial fanatics had a field

114   day, imagining the desperate boy searching the coffins of his relatives for food, but the truth must have been that grave robbers had broken into the vault, which belonged to a well-known wealthy family, in search of valuables. Other causes of presumed premature burial may have been due to the interference of linen thieves out to steal the shrouds or of body snatchers, who were after fresh bodies for dissection and dumped the putrefied ones back into the coffins without ceremony.

Thus, a fair proportion of the old cases of presumed premature burial may well have a natural explanation, and there is no doubt that contemporary writers greatly exaggerated the risk. Some of their accounts are convincing enough, however, and it is inevitable that a few cases are genuine. During cholera epidemics, it was the practice to bury the dead hastily to try to stop the contagion from being passed on; furthermore, patients in the so-called cold stage may look as if they were dead without losing their chances of eventually surviving the infection. It is no coincidence that the famous painting *Inhumation précipitée*, by the Belgian artist Antoine Wiertz, portrays a cholera victim coming to life in his coffin. The sheer number of people dying each year all over the world and the sometimes uncertain methods of declaring people dead in developing countries would indeed imply that even today people may be at risk of being prematurely buried.

In the late twentieth century, there is little trace of the old fear of premature burial; it has been superseded by the modern medical nightmare of being kept alive indefinitely in a comatose state, by means of a respirator and parenteral nutrition. The horror of a living death is still with us, although it

Fig. 4. Inhumation précipitée by the Belgian artist Antoine Wiertz (1854). Reproduced by kind permission of the Musées royaux des Beaux-Arts, Brussels.

has changed its face. It is true, however, that the modern methods of electro-
cardiography and electroencephalography, which are usually sure indicators
of loss of heart and brain activity, may fail in cases of severe exposure to cold
or overdoses of barbiturates or other drugs. In 1970 the French journal *La
Presse Médicale* detailed three cases of people who had attempted suicide by
taking high doses of barbiturates. Despite careful monitoring of their life
signs, they had been erroneously diagnosed as dead; one of them had actually
recovered while laid out in his coffin. In 1988 a man in Mons in Belgium was
diagnosed as being both heart and brain dead after severe exposure to cold;
he revived after long having remained in a comatose state with a very low
heart rate. The combination of hypothermia and a drug overdose has sometimes confused medical attendants and led to a premature declaration of
death. In 1986 a twenty-seven-year-old drug abuser was found lifeless in the
woods near Reigate. According to an article in the *Daily Mail*, he was certi-
fied dead at the New East Surrey Hospital, Redhill. In the morgue a techni-
cian heard spluttering sounds from the alleged corpse, but attempts at resuscitation were vain: the man died thirty-six hours later. There are several
similar newspaper accounts of people waking up in the mortuary or even on
the autopsy table after taking an overdose of drugs or alcohol while outdoors
in a cold climate, and being erroneously declared dead.

I myself in 1988 had occasion to observe a woman arriving at the emergency ward in a comatose state with an extremely low body temperature.
She was an old alcoholic who had, on New Year's Eve, celebrated the occasion with several bottles of vodka, seated on her balcony in freezing cold
weather. She fell asleep out there and was well nigh dead when finally
brought in; her body temperature was far below that considered, at the time,
consistent with human life. The body was cold and deathlike, and there was
no pulse to be felt; the policemen considered her to have expired, and her
skinhead nephew mourned her as dead. The electrocardiogram, however,
revealed a heart rate of twenty and waning! Through gradual extracorporeal
warming of the blood by means of a heart-lung machine her life was saved.
This treatment was remarkably successful, and the woman recovered without any ill effects; her marked obesity, isolating the inner organs from the
cold, was probably to be thanked for her survival. When I told her, in no
uncertain terms, how she had cheated the Grim Reaper, hoping that this
gruesome experience would wean her from the bottle and prevent her death
from alcoholic cirrhosis of the liver, she declared that she had indeed learned
a lesson. Never again would she drink heavily — while seated outside in sub-
zero temperatures!

Today, premature burial has joined the ranks of modern newspaper myths, case reports being recounted in the "funny pages" of the tabloid newspapers, along with stories of spontaneous human combustion, snakes in the stomach, crop circles, and rains of living toads. According to one popular story, when a wealthy American is taken to his funeral in a luxury hearse, a tire blows, and the hearse skids through the front window of another undertaker's parlor. Its doors burst open and the coffin is flung into the premises. The hearse driver and the mourners are astonished to see the alleged corpse walking out through the shattered glass, dressed in his white burial robes. Another, even less likely urban legend is that of the apparently dead woman who recovers inside her coffin, opens the lid while it is actually being lowered into the grave, jumps out, and runs out into the road, where she is run over and killed by a car. These newspaper stories may be considered the modern equivalents of the Lady and the Ring and the Lecherous Monk, adapted to suit the shallow morality of twentieth-century tabloid readers. Actually, there are shades of both these old stories in the recent distasteful urban legend in which the corpse of a young girl revives after she is raped by a perverted grave robber; it is usually added that the girl's parents did not press charges against the rapist, since they thanked him for their daughter's life. An old classic is the corpse sitting up in its coffin after it has been struck by lightning; this urban legend has been repeated at regular intervals since the mid-nineteenth century. In another, perhaps more plausible newspaper story, an old woman in New York, who has taken a couple of stiff brandies before going to bed, loses her bearings in the corridor of the senior citizen's home where she lives, and goes into the wrong room. She undresses and lies down in the bed after pushing a corpse, which had been left in the bed, aside. When the undertakers arrive, they put the old lady in a body bag, and she awakens from her booze-up in a coffin. Fortunately, one of the undertaker's men notices a movement before the presumed corpse is carried out to the graveyard.

🐦 The German sentimental novelist Jean Paul Richter, who had quite a large following in his day, was one of the first to use the uncertainty of the signs of death as a plot device. In his novel *Siebenkäs*, published in 1796, the main character, a young lawyer, plays dead to get away from his wife. The doctor is completely fooled by this simple trick, and Herr Siebenkäs is declared dead, put in a coffin, and buried, to be dug up and rescued by a friend in the next chapter. That famous Irish master of the horror story, Joseph Sheridan Le Fanu, took the perils of being prematurely buried much

Fig. 5. The apparently dead lady awakens. Another drawing from *The Uncertainty of the Signs of Death*, the pirated English version of Bruhier's book, published in Dublin in 1746. Reproduced by permission of the British Library, London.

more seriously. In his short story "The Room in the Dragon Volant," a wealthy young Englishman falls foul of a band of French impostors. He is duped into drinking poison and falls into catalepsy; the rogues put him in a coffin, intending to have him buried alive, when they are stopped by the French police.

A time-honored literary use of the apparent death theme involves some sorcerer or skillful physician who is in possession of a drug capable of inducing a cataleptic trance, indistinguishable from death itself. After being declared dead, the individual who has taken this drug can return from the tomb in a later chapter to dramatic effect. This theme has been used by Boccacio and in Shakespeare's *Romeo and Juliet*; it also occurs in Alexandre Dumas's *Count of Monte Cristo*. Here, the young Valentine de Villefort, the daughter of the wicked procurator who has deceived Edmond Dantès, is gradually poisoned by her murderous stepmother, who has already killed off a considerable part of the family to secure her own son's inheritance. The omnipotent Monte Cristo has seen through her nefarious plot, and he comes up with a clever, if rather far-fetched, plan of his own: with a powerful antidote, he neutralizes the poison every time it is administered, and finally, he visits the girl at night to inform her of the impending danger. This time, he replaces the poison with a drug of his own, which is capable of inducing catalepsy, with the words: "Whatever may happen, Valentine, do not be alarmed; though you may suffer; though you may lose sight, hearing, consciousness, fear nothing; though you should awake and be ignorant where you are, still do not fear; even though you should find yourself in a sepulchral vault or a coffin." This overblown tirade is, little wonder, greeted with the cry "Alas! Alas! what a fearful extremity!" from the terrified girl. Everything moves smoothly according to Monte Cristo's plan, however: Mlle de Villefort is declared dead by both doctor and coroner, and she is interred in the family vault; the Villefort family is rightly punished, and the distraught count fortunately does not forget to have Valentine rescued from the vault and reunited with her fiancé, after she has lain in her coffin for almost a hundred pages.

The deadhouse of Frankfurt-am-Main was a particularly long-standing one, and it is likely to have been visited by the British novelist Wilkie Collins. The ending of his 1880 novel *Jezebel's Daughter* actually takes place in this sinister location. Unfortunately, this novel was just a potboiler and far inferior to his masterpiece *The Woman in White*; his use of premature burial as a literary device is not less than ludicrous. A certain Mrs. Wagner has died and is taken to the deadhouse for three days of observation before the doctor can sign a certificate that putrefaction has occurred. A foundling boy named Jack, to whom Mrs. Wagner has shown much kindness, refuses to leave the body, believing her still to be alive, although the doctor shows him the brass thimbles fitted to the corpse's fingers, and explains the workings of the machinery. Together with an old night watchman, Jack later drinks prodigious amounts of brandy, and they sing a song about a former watchman going mad in the deadhouse:

> *The moon was shining, cold and bright,*
> *In the Frankfort Deadhouse, on New Year's night*
> *And I was the watchman, left alone,*
> *While the rest to feast and dance were gone;*
> *I envied them their lot, and cursed my own —*
> *Poor me!*

Unbeknown to either of these sottish fellows, a certain Madame Fontaine has crept into the deadhouse; she has in fact poisoned Mrs. Wagner, and comes to gloat at the sight of her remains. Even the steely nerves of this murderous hag get a severe shock, however, when she sees the corpse: "There, grand and still, lay her murderous work! There, ghostly white on the ground of the black robe, were the rigid hands, topped by the hideous machinery which was to betray them, if they trembled under the mysterious return of life." She seeks comfort with Jack and his dead-drunk companion, but they scoff and deride her. They sing:

> *Any company's better than none, I said;*
> *If I can't have the living, I'll have the dead!*
> *In one terrific moment more,*
> *The corpse-bell rang at each cell door!*

While they are singing, the bell at Mrs. Wagner's cell rings time after time; only Madame Fontaine hears it, since the other two are too drunk. At long last, the apparently dead lady manages to make her resurrection from the dead known to her sottish attendants, and Jack is reunited with his apparently dead mistress. Madame Fontaine, who inadvertently drinks a glass of her own poison during these proceedings, dies in the agonies considered fitting for a Victorian villainess.

The writer with most premature burials per page must be Edgar Allan Poe, whose unwholesome fascination for this subject is apparent to every

120   devotee of his horror stories. What reader of Poe does not recall Lady Madeline Usher's dramatic return from the tomb in the masterly *Fall of the House of Usher*, or the macabre discovery of the living corpse's extracted teeth in *Berenice*? Popular books on premature burial were readily available in the United States, and it is likely that Poe had read some of them, as is hinted in one of his less widely known stories, "The Premature Burial," which was published in 1844. The direct point of departure for this, for Poe, rather vapid and ineptly plotted story, seems to have been an article in the *Columbian Lady's and Gentleman's Magazine*, a periodical to which Poe was an occasional contributor, about a "life-preserving coffin" exhibited at the American Institute in New York. This coffin, luxuriously made, with soft stuffing and a pillow for the head, was intended for vault burials and equipped with springs and levers on the inside, which would make the lid fly open at the slightest movement of its occupant. It was recommended that the family vault also be equipped with a loud bell, with which to call attention; the bell rope was fastened to the corpse's hands through an opening in the coffin. The magazine also contained a poem in praise of the coffin, written by a certain Mrs. Seba Smith, ending with the lines:

> *And there they saw their daughter,*
> *As the moonbeams on her fell,*
> *In her narrow coffin sitting,*
> *Ringing that solemn bell.*

Fig. 6. A Swiss lady awakens in her coffin just as it is prepared for burial; her relatives rush away, believing that she is a ghost. A drawing in the author's collection, marked "Basler Hinkender Bote, 1822."

The protagonist of Poe's *Premature Burial* is a young man who suffers from a strange neurological disease that makes him prone to epileptic seizures, after which he often lapses into a state of deep lethargy. He is an avid reader of stories about apparently dead people awakening in the tomb and suffering the most dreadful tortures. As a precaution, he buys a life-preserving coffin, and he always carries with him written instructions that he is under no circumstances to be hastily buried, should he suddenly collapse in the street. Finally it happens that he awakes in absolute darkness, lying in a musty-smelling, narrow area, surrounded by wooden walls. His jaws are bound up like those of a corpse, and with horror, he understands that he has been buried alive. In vain he searches for his safety equipment inside the coffin, but he finds nothing but the unyielding walls of his narrow house; he realizes that he must have been buried by strangers, "nailed up in some common coffin — and thrust, deep, deep, and for ever, into some ordinary and nameless *grave*." His wild cry of terror and anguish calls the attention of some sailors, however, and it turns out that he has actually been lying in a narrow bunk on board a trawler; the musty smell comes from the ship's cargo, and the cloth around his jaws is his own necktie, which he had been using as a nightcap. This nightmare experience drives out his fears of being prematurely buried, and transforms him into a saner and healthier man. He burns his medical books on the uncertainty of the signs of death and thereafter reads "no *Night Thoughts*, no fustian about churchyards — no bugaboo tales — *such as this*."

Mary Toft,
the Rabbit Breeder

❧ *Poor Mary Toft in ignorance was bred.*
And ne'er once betray'd a deep designing head.
Ne'er seem'd cut out for plots: Yet never did wife
Like her impose so grossly on Man Midwife.
Who scorning Reason Common Sence and Nature
Plac'd all their faith in such a Stupid Creature.

"The Doctors in Labour"

IN THE TRANQUIL VILLAGE OF GODALMING, situated near Guildford in the county of Surrey, lived the heroine of this strange tale, the poor peasant woman Mary Toft. In 1726, the year of her fame, she was about twenty-five years old. She had been married to her husband Joshua, who was an unsuccessful journeyman clothier, for six years; they had three living children. Mary Toft was described as a short, stoutish woman with coarse features, completely illiterate, and of a "stupid and sullen temper." In April 1726 when Mary was five weeks gone with child, she saw a rabbit jump up before her when she was busy weeding in the fields; she vainly pursued it, but it got away. This event gave her a peculiar craving for rabbit meat, which she could not satisfy because of poverty. Night and day she dreamed of delicious rabbit stew and jugged rabbit. Apparently, her craving had a most sinister influence on her organs of repro-

duction, as expressed in the poem "The Doctors in Labour," one of several   *123*
lyrical celebrations of her bizarre exploits:

> *The Effect was strange — Blest is the Womb that's barren*
> *For that can ne'er be made a Coney warren.*

Four months after seeing the rabbit in the fields. Mary suffered a miscarriage, as it seemed; she was suddenly struck by colicky pains and "a large
lump of flesh" came away. Three weeks later, a similar incident took place,
but her symptoms of pregnancy remained. On the night of September 27,
she was taken very ill and sent for her mother-in-law, who was a midwife.
Now, she "voided somewhat, which she took to be the Lights and Guts of a
Pig." At this stage, Mr. John Howard, the Man Midwife of Guildford, was
consulted. Of this gentleman's earlier career, very little is known, except that

Fig. 1. Mary Toft's
portrait, painted in
1726 by the artist
John Laguerre. From
a print in the author's
collection.

124   he had practiced midwifery for more than thirty years and was well known as a reliable obstetrician. With increasing horror and surprise, Howard delivered Mary Toft of several other parts of a pig. These obstetrical miracles ceased, at least temporarily, and Mary was "churched" a fortnight later. Howard had not seen the last of his strange patient, however. In early October, she gave birth to a rabbit with a cat's paws and head, and then she bred one rabbit after the other. The animals were always dead, and sometimes even cut in two or three parts.

> *Help help good people — fetch another Neighbour*
> *Her pains are strong — She'll quickly fall in Labour*
> *Here Doctor — here good women — help to hold her*
> *Poor thing, she faints — take care — you hurt her shoulder*
> *Bless me! What's this you've brought to town? — O Mary*
> *Three Cats legs and a Coney Skin all hairy.*

Howard was eager to spread the word about these strange births, and he described them vividly in several letters to noblemen and distinguished medical practitioners in London. Every delivery of rabbits was preceded by a strange, pulsating movement in Mary's lower abdomen, which Howard presumed to be the young rabbits jumping in the Fallopian tubes and uterus while making their way toward freedom. Among the country people, word about the miracle in Godalming spread like wildfire, and large crowds went to see Mary and her strange "children," which Howard had preserved in a row of bottles. When eleven rabbits had been born in this way, Howard wrote to the king's secretary: "As soon as the eleventh Rabbet was taken away, up leapt the twelfth Rabbet, which is now leaping. If you have any curious person that is pleased to come Post, may see another leap in her uterus and shall take it from her if he pleases."

The Prince of Wales was astounded by this unheard-of obstetric phenomenon, and he dispatched his secretary, the Hon. Mr. Samuel Molyneux, to find out if it was true. King George I ordered the court anatomist, Mr. Nathaniel St. André, to accompany him. Molyneux was a talented scientific amateur and a fellow of the Royal Society; a year earlier, he had constructed one of England's first telescopes in his home near Kew. He lacked experience in medical matters, however, and let the court anatomist take command on their strange mission to Guildford, where Howard had taken Mary after the birth of the ninth rabbit.

In spite of his distinguished title, Nathaniel St. André was of plebeian origin. He was born in Switzerland in 1680, and had accompanied a Jewish family to London as a foot page. Later, he earned a meager living as a teacher of German, French, and dancing. To augment his failing income, he also gave lessons in fencing, although he was not, apparently, much of a swordsman. Soon, he was badly injured by one of his pupils and had to be carried away to a surgical practitioner. St. André was so impressed by the surgeon's apparent wealth that after he recovered from his wound, he exchanged the rapier for the scalpel. After being apprenticed to a London surgeon, he later set up a lucrative practice, as well as acting as local surgeon to the Westminster Hospital Dispensary. Although he never bothered to take any medical degree, St. André was soon a rising star in the London medical world. This was the time of the first Hanoverian king, and George I had brought many German courtiers with him to England. The clever dancing master, fluent in German and a shameless toady, soon won the favor of the courtiers who couldn't learn

Fig. 2. The "rabbit doctor," Nathaniel St. André, a portrait by an unknown artist. From a print in the author's collection.

English, or hadn't bothered to try. In 1723 he was appointed court anatomist to King George, and three years later, his career reached its highest peak when the king consulted him in person, for what ailment is not known. It is unlikely that it was anything serious, since St. André obviously did quite well, and the patient was cured. The king rewarded him with a sword, which he took from his own side for the purpose.

The great success of this foreign upstart made him unpopular in London's medical world, where his meteoric career aroused the envy of those less fortunate. According to the *Dictionary of National Biography*, even St. André's partisans had to allow that he was ignorant, foul-mouthed, and lecherous. In 1725 he claimed that he had been decoyed and poisoned by the hired hand of some adversary, but although the Privy Council promised a reward for the capture of the alleged poisoner, no offender was found. Many people thought that the whole thing was a trick by St. André to get his name before the public, since he was very keen to advertise his practice. It is no coincidence that he was the man sent to Guildford to examine the rabbit breeder: he would indeed achieve public fame from this ill-fated expedition, but not in the way he had calculated.

Now Mary struggles with a second Pain
The Doctor now atends her throne again
But ah too late — Impatient of delay
Bunn thro' his Burrough works himself a way
Tho not so slily but the Doctor spies him
And follows with design t'Anatomize him.

As soon as St. André and Molyneux reached John Howard's house in Guildford, where Mary lodged as a resident patient, the excited man midwife came running out to tell them that the rabbit breeder was at that instant in labor with the fifteenth rabbit. Some minutes later, the court anatomist delivered her of the trunk of a rabbit of about four months' growth stripped of its skin but containing the heart, lungs, and diaphragm. When Molyneux and St. André examined the rabbit, they found that the lungs floated in water; the obvious implication — that the rabbit had breathed air before its death — escaped the gullible court anatomist and his puzzled companion. St. André was very much impressed by this strange birth, and his opinion was rather that these preternatural "rabbets" did not follow the ordinary laws of physiology. Mary herself seemed very much relieved and cheerful, and sat down in a chair before the fire, while the proud rabbit man midwife took his two visitors on a guided tour of his anatomical museum, which contained the previous fourteen rabbits, all preserved in jars of alcohol. The first one had three cat's paws, but the others all seemed like normal young rabbits, of two to four months' growth. They had all been delivered in several pieces; the ingenious Mr. Howard believed that the animals had been crushed into pieces by the powerful contractions of the uterus. St. André examined several of the rabbits, without finding anything abnormal. Two hours later, Mary again fell into labor, and the nurse delivered the lower body of a male rabbit, also stripped of its skin and perfectly fitting the part delivered earlier. On their return, the two royal emissaries "anatomized" this specimen, finding in the rectum several pellets of a substance "much of the same Colour and Consistence as the common Dung of a Rabbet"—another piece of evidence that the rabbits were preternatural!

Later the same evening, Mary was once more struck by violent labor pains, and it took five people to confine her to the chair. First, she gave birth to the skin of the rabbit, rolled together like a ball, and finally the animal's head with the "furr" on. St. André examined her abdomen closely, noting some irregularities at the site of one of her ovaries. Taking into account the pulsating movement in her belly, the man of science presumed that the rabbits were formed in the ovaries and made their way to freedom by jumping down the Fallopian tubes. When he left Guildford, St. André took several rabbits with him to show them to the king and the Prince of Wales. With great expedition, he wrote a complete account of the Guildford miracle titled *A Short Narrative of an Extraordinary Delivery of Rabbets*. The Hon. Mr. Molyneux added a rather cautious postscript stating that he had not observed any signs of fraud. Like the ungenerous frontline scientists of the present day, always eager to secure priority in discoveries in progress, St. André promised that "The Account of the Delivery of the eighteenth Rabbet, shall be Published by way of an Appendix to this Account."

> *The Doctors here and Midwives all consult*
> *If 'tis a foetus Rabbit or adult*
> *When up the learned Merry Andrew starts*
> *This Animal (quoth he) in all its parts*
> *Does with a Natural Rabbit well agree*
> *And therefore it must Praeternatural be.*

Before a large audiences of Britain's finest, including the king, the prince, and many of the foremost nobility, St. André made an anatomical demonstration of the first, third, fifth, and ninth rabbits on November 26. Those present seemed to be convinced by the court anatomist's learned harangues, and the king ordered that Mary was to be brought to London for strict supervision. The future rabbits were thus to be born in the heart of the metropolis. This plan of action was also recommended by a German surgeon named Cyriacus Ahlers, who had been sent by the King to examine Mary the week before. Ahlers, who does not seem to have been a much more skillful practitioner than St. André, had delivered the sixteenth rabbit with his own hands. When he asked Howard why the rabbits were always skinned, the fast-witted man midwife answered that this was obviously the effect of the strong pressure of the womb toward the pubic bone. This theory was accepted by the dense German surgeon, but he did find it peculiar that although Mary had "began to squall and roar" during her labors, she had pressed her knees together as if she did not want to drop her preternatural fetus prematurely. Furthermore, when Howard and Ahlers cracked some jokes while Mary was in labor, the German was astonished that "the Patient laugh'd very heartily with us"; the resourceful Mr. Howard once more saved the day by emphasizing her very strong constitution. Ahlers wanted to remain in Guildford to await further miracles, but John Howard did not allow him to stay, since the fumble-fisted German, who lacked obstetrical training, had hurt Mary when he tried to extract the rabbit. Before leaving, Ahlers gave Mary a sum of money, and he also put out promises of a royal pension when she had been taken to London. The shameless John Howard wanted to secure a similar pension for himself, as a suitable reward for England's foremost rabbit man midwife.

When St. André's pamphlet about the rabbit breeder reached the bookstalls of London, it instantaneously became a best seller. Rumors had been buzzing for several weeks about the strange things afoot in Guildford. On December 2 Lord Hervey wrote to Henry Fox: "There is a thing that employs everybody's tongue at present, which is a woman brought out of Surrey who had brought forth seventeen rabbits, and has been these three days in labour of the eighteenth. I know you laugh now, and think I joke; but the fact as reported and attested by St. André the surgeon (who swears he has delivered her of five) is something that really staggers one." Alexander Pope wrote to his friend John Caryll to inquire what faith he had in the rabbit breeder, since all London was divided into factions about her, and Lord Onslow wrote to Sir Hans Sloane, lamenting that "the affair had almost alarmed England, and persuaded several people of sound judgment that it was true." According to a letter still kept in the Sloane papers at the British Museum's department of manuscripts, St. André also wrote to Sir Hans Sloane, inviting him to come and see Mary and her strange children.

Indeed, the London public spoke of little else than the rabbit breeder and her exploits. Religious people suspected that witchcraft was involved and that Mary was really a rabbit in the shape of a woman. Individuals of more worldly bent hinted that the rabbit breeding was due to a "criminal conversation" between Mary and a male rabbit. One pamphlet writer even intimated that she had kept a large buck rabbit as a pet and done it various unspecified "acts of kindness." Some know-it-alls blandly asserted that she was an impostor, having allowed a doe rabbit to kindle within her.

While the ignorant public thus spent their time in idle speculation, the men of science searched for precedents for the Guildford miracle. Although early eighteenth-century medical science was unwilling to accept that women might give birth to animals, the old annals of strange and macabre events listed quite a few obstetrical mishaps similar to the Guildford wonder. Had not Pliny written of the Roman lady Alcippa, who gave birth to an elephant, and did not Conradus Lycosthenes mention an Italian woman delivered of a kitten and another of a young dog? Ole Worm, the famous Danish anatomist, had described a strange case from Norway in his monumental *Museum Wormianum*: in 1638 a peasant woman had laid two eggs resembling common hens' eggs. One of these "magical eggs" was kept in the king of Denmark's private museum in Copenhagen for many years, before being sold at a public auction in 1824. Thomas Bartholin, Worm's foremost pupil, incorporated a strange tale of a noblewoman from Elsinore into his collection of anatomical curiosities: she gave birth to a large rat, which dashed around the room like a fury, dodging the futile attempts of the midwives to catch or kill it.

One of the few London obstetricians with personal experience of women who bred animals was Dr. John Maubray, who had discussed this subject in his book *The Female Physician*. He believed that Dutch women were disposed to bring forth an evil-looking little animal, which they called *de suyger* or *sooterkin*, particularly if they spent much time before their hot stoves. The sooterkin sucked all the infant's nourishment, like a leech, and the child was dead and desiccated when it was born in such unpleasant company. Dr. Maubray even claimed that he had seen and delivered a sooterkin when he was traveling on a ferry from Harlingen to Amsterdam, and a woman fell into labor on board. The creature, he reported, was "the likest of anything in Shape and Size to a *Moodiwarp* (mole); having a hooked Snout, fiery sparkling Eyes, a long round Neck, and an acuminated short Tail, of an

130 agility of Feet. At the first sight of the World's Light, it commonly Yells
and Shrieks fearfully; and seeking for a lurking Hole, runs up and down
like a little Dæmon." Maubray's professional colleagues received this sensational account with scorn, and the London wags put out a pamphlet titled
The Sooterkin Dissected, proving to their satisfaction that neither God nor the
Devil could have made an animal of this kind and that only the most stupid
and impudent would describe such a beast. Among simple village practitioners, Maubray's book was doubtless received with more enthusiasm.

> *Now to the Bagnio flock the Town & Court,*
> *T'improve their judgments some and some for sport,*
> *They're wellcome all to Mary — all that will*
> *May in her Warren for a Rabbit feel.*
> *But Moll take care they don't ye Trick discover*
> *For then thy Merry days will all be over.*

On November 29 Mary was installed at Lacy's Bagnio in Leicester Fields,
a public bath that women sometimes used for their lying-in. St. André and
Howard were received with great deference, and the public held the heroine
herself in superstitious awe. It was generally hoped that the eighteenth rabbit
would soon make its appearance. Many high noblemen had applied for seats
at the Bagnio in advance, as if it were to be a fashionable stage premiere.
Some of them no doubt were motivated by scientific curiosity; others were
lechers and rakes who wanted to examine the marvelous "coney-warren"
with their own hands. Lord Hervey wrote that "every Creature in town both
Men & Women have been to see & feel her." The unhappiest men in London
were the warreners and poulterers, whose business reached an all-time low
after the Guildford miracle became public knowledge. The reason for this
decline was not only that many people felt it to be distasteful to eat rabbits
after the last weeks' startling revelations but that a law from the time of
Queen Elizabeth forbade the eating of anything that might be borne by a
woman. The director of the play *The Necromancer*, which was acted at the
Theatre Royal, Lincoln's Inn Fields, introduced a new scene in which Har-
lequin, dressed as a woman, pretended to be in labor and was delivered first
of a large pig, then of a sooterkin and several other animals, to the great
amusement of the audience.

At Lacy's Bagnio, several skilled physicians and men midwives joined
the team of medical men in attention. The Princess of Wales had recruited

Sir Richard Manningham, FRS, one of the leading obstetricians of his time, who is mentioned in Laurence Sterne's *Tristram Shandy*. His colleague as a rabbit obstetrician was the famous surgeon James Douglas, a specialist in the anatomy of the female pelvic organs. The notorious "Sooterkin Doctor" John Maubray also appeared at the Bagnio, probably much pleased that his theories seemed to be being vindicated and happy to be consorting with St. André, the man of the day, and his companion John Howard. Maubray

The Surrey Rabbet-Breeder *here behold,*
Imposture greater than appear'd of Old,
Mountains in Foreign Climes w. Labour gro
And One small Mouse if dire Surprize atton'd,
But here by Mounsieur's *Art at last reveal'd*
Lye Seventeen Coney's *in one——conceal'd,*
Poor Mary Toft *the Tool is only made,*
While her Assistants *do perform their Trade*

The
SURREY-WONDER
an Anatomical Farce
as it was Dissected
at if Theatre Royal
Lincolns Inn Fields

Pr 6 d

Yet Act by turns their Parts so very ill of skill
That their own Hands have Sign'd their want
Hence let S. t Wretches *learn if e'er they Aim*
T 'establish Fraud, *their Character's if Game,*
The Hunt once up, Mankind will Cry 'em down
And make such Quacks *the Sport of Court & Town.*
The Eighteen. Birth *whenever it appears,*
'Tis hop'd will bring forth Pillory *and* Ears ,

Sold att the grocers Shop the corner of Oxindon Street in Coventry Street up one payr of stairs Pickadilly

Fig. 3. The Surrey Wonder by James Vertue, another drawing of Mary and her rabbit children, which was confiscated by the crown, since one of the characters behind the rabbit breeder resembled the Prince of Wales. It is likely that the drawing was inspired by the enactment of the obstetrical farce at the Theatre Royal, Lincoln's Inn Fields. Reproduced by kind permission of the British Museum, © British Museum.

could now claim a resounding victory over his adversaries in the sooterkin debate: it was indeed possible for a woman to give birth to animals. Most of the medical men believed that Mary's rabbit breeding was due to the force of the imagination and that her wanton longing for rabbit meat had transformed her unborn child into a whole warren of these animals. Dr. Maubray had indeed warned about this possibility in his *Female Physician*, advising women not to "please themselves so much in playing with Dogs, Squirrels, Apes, &c.," lest their child resemble its mother's favorite pet.

Manningham and Douglas were very skeptical of the whole business, however, and Douglas thought human conception of a rabbit was about as likely as rabbit conception of a human child. With their assistants, who were elected from the foremost nobility — only dukes were put on the roll for emergency duty — they took turns to watch the rabbit breeder day and night to make sure that she was not supplied with fresh rabbits from outside.

When Sir Richard Manningham had been sent by the king to fetch Mary to London, he had inspected a specimen alleged to be the placenta of the last rabbit. Although John Howard had accepted it, Manningham bluntly commented that it looked more like half a hog's bladder. At this remark, Mary started to weep profusely, exacerbating Sir Richard's suspicions. He asked for a hog's bladder, and the resourceful rabbit man midwife of Guildford was able to produce one from his museum; the two specimens were exactly alike, and both had "the same strong *urinous* Smell peculiar to a Hog's bladder." Manningham brought the bladder with him to London, and since Mary's labor pains had miraculously ceased as soon as she entered the Bagnio, this unprepossessing specimen served as cynosure; it was handed around and examined until it was almost worn out. Manningham and Douglas wanted to announce in public that they suspected a trick, but St. André and Howard persuaded them to wait a few more days. They predicted that Mary would soon breed new rabbits. St. André declared that he did not find the birth of half a hog's bladder any more remarkable than Mary's previous breeding of seventeen preternatural rabbits, but even his own supporters doubted the wisdom of this statement. St. André managed to retain his *sang-froid*, and he assured the noblemen and courtiers that great wonders were still to be expected from his vulgar protégée. He bullied and insulted the medical men who wanted access to the rabbit breeder, preferring an audience of peers and lecherous men-about-town for his pompous lectures about Mary's strange births. Once, he even turned Dr. Douglas himself away, and this gentleman indignantly dismissed the whole affair, commenting as he left that "some new monster was a-breeding."

꩜

'Tis an unhappiness to be Lamented
That people ne'er know when to be contented,
Had breeding seventeen Rabbits satisfied
Poor Mary Toft the Plot had still been hid;
But fond to make the Number up a score,
The prying World the secret did explore.

On December 4 the obstetrical farce at the Bagnio accelerated once more. Mary suffered violent labor pains, and Dr. Maubray and the German obstetrician Dr. Limborsch agreed with Sir Richard Manningham that "something would soon issue from the *uterus*." St. André and Howard were much enlivened by the good news and bragged about the great things to come. That same evening, disaster struck. Thomas Howard, porter at the Bagnio, went before Sir Thomas Clarges, justice of the peace, to state that Mary Toft had bribed him to bring a rabbit to her. Under interrogation Mary confessed to the attempted bribery of the blabbering domestic; she stubbornly maintained that she had wanted the animal only to cook and eat it and that she was pregnant with more rabbits. Two days later, Mary was again interrogated, and Manningham threatened her with a horrible operation to explore her pelvic organs if she did not confess. Some wags claimed that he had promised to send a chimney sweep's boy up her Fallopian tubes to explore them! At any rate, this dire threat frightened the wretched woman a great deal, and she promised a public confession the next day. This was taken down before Sir Richard Manningham, James Douglas, Lord Baltimore, and the duke of Montague, and it is still preserved among the Douglas files at the Hunterian Library in Glasgow. Mary had put quartered rabbit bodies into a special "hare pocket" inside her skirt and had introduced them into her vagina whenever she was not closely watched. Then she would fall into "labor" in a theatrical way and surrender the unsavory trophy to her puzzled obstetrical attendants. Mary was agreeably surprised that it took so little skill to fool men such as Howard and St. André. When she was taken to London, it was of course more difficult for her to obtain rabbit parts, since she had not had the wit to bring some from Guildford. When the medical men had grown impatient, she had made the desperate attempt to bribe the porter which brought about her ruin.

Sir Richard Manningham published an "Exact Diary" of his attendance on Mary Toft, in which the extraordinary imposture was unveiled to the public. He emphasized his own part in detecting the woman's tricks, which greatly offended James Douglas, who soon published his defense, *An Adver-tisement occasion'd by Some Passages in Sir R. Manningham's Diary*. Lord Onslow, a prominent justice of the peace, had gone to Guildford to inquire into the rabbit trade there; several rabbit mongers and other witnesses deposed that Joshua Toft had been a frequent purchaser of young rabbits at the time of his wife's lying-in. Cyriacus Ahlers feared that his part in the bizarre proceedings in Guildford would jeopardize his promising career at court, since St. André had published some affidavits from John Howard and others in the second edition of his notorious rabbit pamphlet, alleging that Ahlers had also been convinced that Mary really bred rabbits and that he had promised her a pension from the king. Ahlers published a pamphlet of his own, *Some Observations concerning the Woman of Godlyman in Surrey*. He considered John Howard little better than a quack and impostor; the Guildford man midwife had examined his patient when he was dressed only in a nightgown and had assumed an unusual, not to say indecent, posture while "touching" her. Cyriacus Ahlers did not dwell on his own doings in Guildford and asserted that he had suspected foul play from the beginning. He emphasized that one of the newborn rabbit's pellets of dung had been proved to contain hay, straw, and corn. Ahlers also hinted that Howard had been a party to the plot from an early date, and this is by no means unlikely; the rabbit man midwife of Guildford may well have been impressed by the ease with which Mary tricked the great medical men from London, and encouraged by the vague promise of money from both St. André and Ahlers. It is certain that he, too, was prosecuted for his part in the affair.

So great was public interest in the rabbit business that Londoners talked of little else during the winter months of 1726, and a declaration of war or a second Gunpowder Plot would hardly have been noticed. There was a great upsurge in interest in the female genital anatomy: terms such as "uterus," "perinaeum," and "Faloppian tubes" (so spelled) were on every man's lips. After the imposture was detected, Grub Street had a field day. The pamphleteers, draftsmen, and writers of lugubrious verse were hard at work in their chambers, and the presses were busy for several months with the production of pamphlets, squibs, broadsides, and ballads about the wretched rabbit breeder. Most of these made poor St. André the butt of their wit, and all London laughed at him.

One scurrilous publication titled *Much Ado about Nothing* was alleged to be Mary's true confession. Some have attributed it to Jonathan Swift, but this vulgar and unladylike tirade, meant to be in Mary's own words (and spelling!) seems beneath the dignity of this great writer. She claimed that

John Howard had always told her that she was "a Woman as had *grate nat-turul parts*, and a *large capassiti*" and that he had "served" her rabbits "tost up skin and all with its eres prickt up"! She described Maubray as "that squab man who cried out a Sooterkin, a Sooterkin," and she declared, "I don't like him at all." Manningham was "an ugly old gentleman in a grate black wig," and the worst of them all, James Douglas, "a fare faced long-nosed Gentil-man, with the neck of a Crain: he was for purformin an oppurashun, and tawkt of making *Insishuns*, and *Cesariums* and the Lord knows what." Jona-than Swift was certainly interested in the Mary Toft affair, and tradition has also accused him of writing another pamphlet, *The Anatomist Dissected*, under the pseudonym Lemuel Gulliver. This satire at St. André's expense also seems rather heavy-handed but perhaps it was written hurriedly, in order to cash in on the rabbit hysteria. Gulliver was particularly incensed that the rabbit affair had adversely affected the sale of his own worthy memoirs, which had previously been quite a best seller. The public, it seems, now pre-ferred the rabbit pamphlets to serious literature.

On December 20 Alexander Pope published his ballad "The Discovery; or, The Squire turn'd Ferret," a bawdy song that was much in favor at rowdy gatherings in the alehouses of the metropolis. Pope suggested that ferrets should have been used to hunt the rabbits out from their hiding places within the unfortunate woman. Unlike the other authors of songs and squibs about the rabbit breeder, Pope was disposed to let St. André off rather easily, since the rabbit doctor had treated him for an injury to the hand two or three months before. Pope's version of the conversation between Molyneux (the Squire) and St. André is particularly funny, especially when Moly-neux gets the idea of using his astronomical telescope in the gynecological examination:

> *If ought within we may descry*
> *By Help of Telescope.*
>
> *The Instrument himself did make,*
> *He rais'd and level'd right*
> *But all about was so opake,*
> *It could not aid his Sight.*
>
> *On Tiptoe then the Squire he stood,*
> *(But first He gave Her Money)*
> *Then reach'd as high as e'er He could,*
> *And cry'd, I feel a* CONY.

Is it alive? St. André cry'd;
It is; I feel it stir.
Is it full grown? the Squire reply'd;
It is; see here's the FUR.

The rabbit was well taken care of, once it was safely born:
He lap'd it in a Linnen Rag,
Then thank'd Her for Her Kindness;
And cram'd it in the Velvet bag
That serves his Royal Highness.

This was one of many allusions to the involvement of the Prince of Wales in the scandal.

Another poem, the illustrated, twelve-stanza "Doctors in Labour" is even more ribald. A verse such as

The Rabbit all day long ran in my head.
At Night I dreamt I had him in my Bed:
Methought he there a Burrough try'd to make
His head I patted and I stroak'd his Back.
My Husband wak'd me and cry'd Moll for Shame
Lett go — What 'twas he meant I need not Name.

must have caused much coarse laughter in the ale-houses! Yet another poem, "The Rabbit-Man-Midwife," was written by Dr. John Arbuthnot, who wanted to satirize Lord Peterborough, a well-known lecher, by naming him as the rabbits' father:

The Doctor search'd both high and low,
And found no Rabbit there.
But peeping nearer cry'd, Soho,
  I'm sure I have found a hare.

They all affirm with one Accord,
  When they had search'd her thorough,
That Bunny's Dad must be a Lord,
  Whose name does end in Burrough.

By the nineteenth century, the many pamphlets, tracts, and poems had become quite valuable, and complete collections bound in rabbit skin were sold for fifteen to twenty guineas. Today, they command high prices at rare

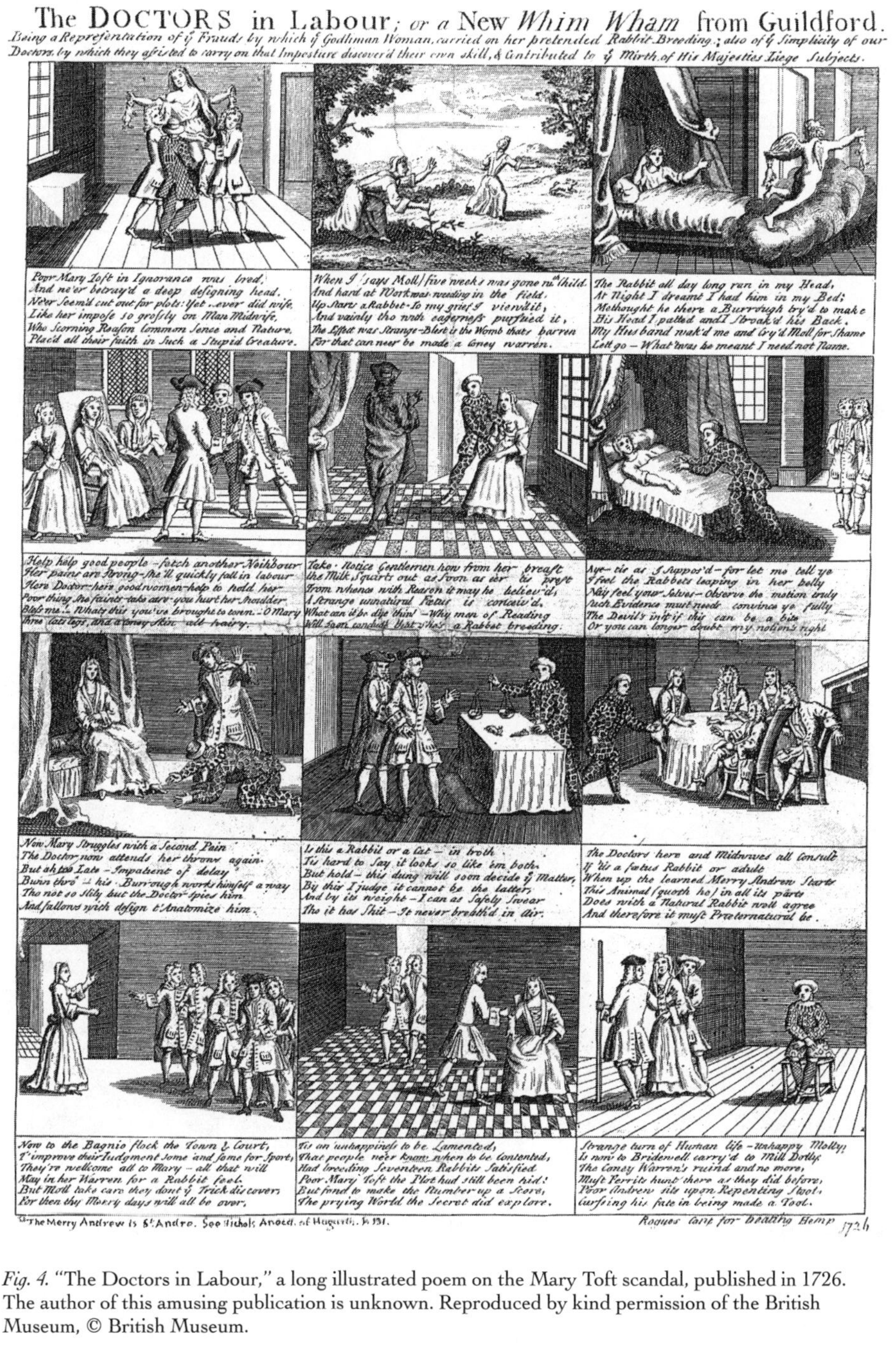

Fig. 4. "The Doctors in Labour," a long illustrated poem on the Mary Toft scandal, published in 1726. The author of this amusing publication is unknown. Reproduced by kind permission of the British Museum, © British Museum.

138   book auctions. One fine collection is kept at the Bodleian Library in Oxford, another at the Waller Library in Upsala. A particularly complete set of pamphlets, poems, articles, and letters, with some unique manuscript additions, is kept at the Library of the Royal Society of Medicine in London. It was gathered by a certain Dr. Combes in the late eighteenth century, and later bought by Dr. Samuel Merriman, who may have added to the collection; he certainly made several manuscript annotations in its pages. Dr. Merriman believed that his collection contained everything published on the rabbit business, but there are at least four more pamphlets and squibs in the matchless collections of the British Library.

At a dinner held by several London doctors some time after the event, it was decided that a caricature drawing was to be made to commemorate it. Everyone present contributed one guinea to this cause, and an engraving was made from a ridiculous sketch by the celebrated William Hogarth. It appeared under the title *Cunicularii; or, The Wise Men of Godliman in Consul-*

Fig. 5. William Hogarth's *Cunicularii* (1726). Reproduced by kind permission of the British Museum, © British Museum.

tation. Depicted, from left to right, are the astonished Joshua Toft, the nurse (probably Mary's sister-in-law Margaret Toft, who was also a party to the imposture), Mary herself as "the Lady in the Straw," Sir Richard Manningham "searching into the depth of things" in his large black wig and obstetrician's smock, Dr. Maubray exclaiming "A Sooterkin!" and St. André dancing with his violin under one arm. At the door, John Howard denies a warrener coming to sell him a rabbit, with the words "It's too big!" In one of Hogarth's later engravings, *Credulity, Superstition, and Fanaticism,* Mary is also depicted, giving birth to a number of rabbits, which are merrily leaping about. At her side is the nail-vomiting Boy of Bilston, who duped several clergymen into believing him bewitched.

❧

Strange turn of Human life — unhappy Molly
Is now to Bridewell carry'd to Mill Dolly

Fig. 6. Hogarth's other drawing featuring Mary Toft, *Credulity, Superstition, and Fanaticism.* A print in the author's collection, dated March 15, 1762.

*The Coney Warren's ruind and no more
Must Ferrits hunt there as they did before,
Poor Andrew sits upon Repenting Stool,
cursing his fate in being made a Fool.*

The Mary Toft scandal displays the London medical world of the early 1700s at its very worst. The doctors were shown up as ignorant, avaricious fools, toadying before the king and the courtiers. Nearly all those involved in the scandal wrote pamphlets afterward to justify their own actions and blacken those of their adversaries. At this time, when many Englishmen resented the Hanoverian king's preference for German-speaking foreigners over native Britons, they took their revenge in a torrent of drawings and squibs ridiculing the Swiss immigrant St. André and the German court physicians as little more than charlatans and the king and the Prince of Wales as gullible fools.

But how ever did Mary get the idea of pretending to breed rabbits? It was by no means the usual way for a young Surrey wench to earn her keep. Some have thought her inspired by Dr. Maubray's ideas about the sooterkin, but Mary was an illiterate countrywoman of relatively feeble intellect, unlikely to tune in to a debate among the London wags about this monstrous little animal. An alternative suggestion is that John Howard, who had certainly read about the sooterkin, planned the whole scheme from the beginning, but in that case, Mary could easily have implicated him in her confession. Furthermore, old wives' tales about women giving birth to animals were probably by no means unknown in the English countryside at this time, and Mary might have found the inspiration for the great drama of her life from some idle tale of a village gossip. Indeed, human beings, or at least that subset which writes and reads that pride and joy of British tabloid journalism, the irrepressible *Sun*, seem to have progressed little since the times of Mary Toft. Some recent headlines from this newspaper are "Woman Gives Birth to 3 Lizard Babies with 12-inch Tails!" and "Woman Gives Birth to Black Sheep!" and even "Chimp Gives Birth to Human Baby!"

The obstetrical farce around poor Mary Toft was not yet played out. On December 9, she was put into Bridewell Prison, charged as a vile cheat and impostor. To prevent any more rabbit breeding, the authorities allowed only the prison keeper's wife to enter her cell, and Joshua Toft was strictly searched whenever he appeared. Worst of all, the prison was open to visitors, and Mary was exhibited by her wardens like an animal in a cage. The London swells paraded outside, ogling her and making obscene jokes, the topic of which was decidedly below the belt; their point was soon worn out, at least for poor Mary, the butt of everyone's laborious wit. Eventually, the king and his counselors decided that they had had more than enough of the rabbit hysteria that still held sway in London. A final bloodthirsty pamphlet claimed to describe Mary's suicide in prison, but she was in fact released after some months and dispatched back to her home in Godalming. The once-famous rabbit breeder probably welcomed her return to obscurity. She was to remain in Godalming for the rest of her life, and little is known of her further career, but there is one curious postscript. The Godalming parish register records the baptism of a daughter, Elizabeth, on February 4, 1727, and notes that this was her "first child after her pretended Rabett-breeding." Thus, she must have been several months' pregnant throughout the whole affair, and her actual pregnancy may partially explain why it took so long to see through her imposture.

Sadly, it seems that Mary Toft's criminal tendencies had not been completely eradicated by her much-publicized brush with the law. In 1740 she was imprisoned for receiving stolen goods. In January 1763 the obituary columns of the London newspapers announced her death among those of peers and statesmen. In 1823, William Clift, the conservator of the Hunterian Museum in London, went to Godalming to search for the rabbit breeder's grave. He found many graves and other remains of the Toft family, but not that of Mary herself. When I made the same pilgrimage in 1994, several gravestones of members of the Toft family, which was quite numerous in Godalming in the seventeenth and eighteenth centuries, could still be identified.

The other leading character in the scandal's dramatis personae, Nathaniel St. André, had as much cause to lament its outcome as Mary herself. Soon after the imposture was discovered, he went to perform his duties at court, but the high-handed courtiers and flunkies there snubbed and insulted him. When St. André was audacious enough to request an audience with the king, he received such an affront from the furious monarch that he never again dared set foot within the royal palace. Many of his former patients left him after he had become the laughingstock of the metropolis, and since he had expensive habits, he was soon in severe penury. Although he retained his title of court anatomist, George I deprived him of both duties and salary.

St. André's companion on the ill-fated expedition to Guildford, Samuel Molyneux, was also much laughed at in town, and he must have been heartily fed up with all the heavy-handed allusions to telescopes, ferrets,

142 Fallopian tubes, and chimney sweeps' boys whenever he attended a dinner party. Nevertheless, he continued his career as a politician and courtier, having married the daughter of the earl of Essex, who was the heiress to a considerable fortune. In 1728, the Hon. Mr. Molyneux was seized by a fit in the House of Commons and died soon after. Since he was one of St. André's remaining patients, there were persistent rumors that the court anatomist had poisoned him. When it became known that St. André and the widow, Lady Elizabeth Molyneux, had left town hastily on the night of her husband's death, London society was aghast.

Although the circumstances were highly mysterious, there was never any direct proof of St. André's guilt, and the Molyneux family was satisfied of his innocence. After her subsequent marriage to St. André, Lady Elizabeth was dismissed from her attendance to Queen Caroline, and the couple had to retire into the country. During his married years, St. André lived an easy life, subsisting on his wife's considerable fortune. According to an annotation in the Merriman volume at the Royal Society of Medicine, his wealth once more ensured him a circle of flatterers, although "his conversation was offensive to modest Ears, and his grey hairs were rendered still more irreverend by repeated acts of untimely Lewdness." On her death, Lady Elizabeth's

Fig. 7. A rare caricature drawing of St. André (Doctor Meagre) receiving a French surgeon sent to investigate the rabbit business, published in 1726. The paintings in the background refer to St. André's earlier fencing career and a stick fight between him and a flower painter outside Slaughter's coffeehouse. Reproduced by kind permission of the British Museum, © British Museum.

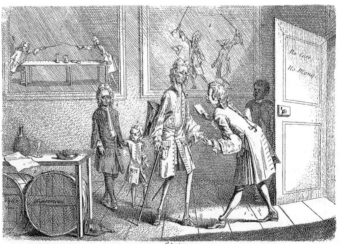

1. Mr. Petit a French Surgeon sent from Paris to Doctor Meagre to take an exact Account from him of ye Præternatural Delivery of Rabbets 2. The most Profound & Learned Doctor Meagre. 3. Doctor Meagre's Son & Heir to his famous Mercurial Pill. 4. Mr. Dipthong Tutor to the Doctors Son and Bosom Counsellor to ye Doctor. 5 The Doctor's Study. 6 A most Heroick Incounter at Cudgels ye happen'd between the Doctor & a Flower Painter at Slaughter's Coffee house. 7 Another Incounter at Swords between the Doctor & a famous French Gramarian, a long Table standing between the Champions. Pri.

money went to her own relatives, and the aging court anatomist was penniless once more. He spent his declining years in an almshouse in Southampton, where, like Thackeray's Barry Lyndon, he sat remembering the days of his youth and fortune. He did not die until 1776 at the age of ninety-six. During the last fifty years of his life, he did not once eat rabbit, and his remaining friends were considerate enough never to serve it whenever he was present at a dinner party.

Maternal Impressions

❧ *Ye pregnant wives, whose Wish it is, and Care,*
To bring your Issue, and to breed it fair,
On what you look, on what you think, beware!

Abbot Claude Quillet, *Callipaedie*

SAINT JEROME AND OTHERS TELL THE TALE of a highborn lady of ancient Greece who stood accused of adultery. The evidence for the prosecution seemed irrefutable; she was white, as was her husband, and yet she had given birth to a distinctly dark-skinned child. Just as the wretched woman was to be sentenced, Hippocrates appeared in court as an expert witness on his own initiative. He pointed out that she had kept a picture of a Moor in her bedroom, and that the child had a fatal resemblance to the individual in the painting. Because the lady had spent much time in her bedchamber, frequently looking at this portrait during pregnancy, she had received a "maternal impression" that had altered the unborn child's shape and skin color. In view of Hippocrates' high standing among his countrymen, none of those present even dared to hint that her penchant for swarthy gentlemen must have gone farther than looking at their portraits; the jurors were convinced, and the lady acquitted, saved by the Father of Medicine. The reaction of her husband has, perhaps fortunately, not been noted for posterity.

The belief in maternal impressions is of great antiquity; it has been traced to ancient India and China, as well as to early African and Asian folklore, to the old Japanese, and to the Eskimos. Its age and distribution appear to equal that of the human race. The Old Testament provides an often-quoted reference. Chapter 30 of Genesis tells how Jacob managed to trick his ungenerous father-in-law, Laban. He agreed with Laban that, as a reasonable salary for his hard work as a shepherd, all newborn speckled or spotted goats and black lambs were to become his property. The parsimonious employer believed that he had made quite a good deal, since these lambs were normally only a small proportion of the herds, but Jacob placed "rods of fresh poplar, and of the almond, and of the plane tree," in which he had peeled white stripes, at the water troughs where the animals mated. This pioneer experiment in genetic engineering was completely successful, for the young conceived over these rods were all speckled or spotted. Jacob built up strong herds of speckled and spotted animals, "and the man increased exceedingly, and had much cattle, and maid-servants, and men-servants, and camels, and asses."

The ancient Greeks and Romans seem to have firmly believed in maternal impressions. It is said of the ugly tyrant Dionysius of Syracuse, that when his queen was pregnant, he ordered her to look at him as little as possible and instead to gaze at a portrait of the hero Jason, which was hung in her bedroom in the hope that the child (evidently the father wanted a boy) would resemble the handsome warrior. Empedocles believed that the features of newborn children often resembled the statues favored by their mothers, and the laws of Lycurgus required Spartan women to look upon statues of heroes such as Castor and Pollux during pregnancy, so that future generations might gain in strength and courage. In Plato's ideal republic, all malformed individuals were to be kept hidden, to prevent imaginative pregnant women from having children exactly like them.

In the early novel *Ethiopica* by Heliodorus, written in the third century A.D., the heroine Chariclea is a white-skinned princess, the daughter of the Moorish king and queen of Ethiopia. A portrait of Andromeda, hung in the royal bedchamber, had caught the queen's fancy to a marked degree. Doubting, however, that the jealous king would share her faith in maternal impression, she has Chariclea adopted away and a black child put in the royal crib in her place. Reaching adulthood, Chariclea returns to Ethiopia after a series of hair-raising adventures, but only to be captured and brought before the throne in chains to be sacrificed to the gods. Fortunately, her unnatural likeness to the portrait of Andromeda is recognized by the astounded queen, who manages to secure a happy ending to this remarkable novel by explaining the whole story to her husband's satisfaction.

Many early chroniclers quote the story of Hippocrates and the adulterous Greek lady as evidence that the Father of Medicine himself believed in maternal impressions, but this story is a later invention and cannot be found in the Hippocratic writings. His only mention of the belief is in the treatise *De superfoetatione*: if the pregnant woman has a desire to eat coal and earth, the child will carry a black mark on its head. Several commentators doubt, however, that this treatise was really written by Hippocrates.

The ancient Romans were firm believers in maternal impressions, and some writers even maintained that all variations in the child's appearance, compared with that of its parents, could be explained in this way. In Latin birthmarks were called *naevi materni*, or "mother's spots," and they were said to originate in some fright suffered by the pregnant woman. Pliny explained the greater variability in human external appearance, compared with that of various animals, by the greater power of imagination of humankind. Nevertheless, it was considered an undisputed fact that animals could also receive maternal impressions. In his *Cynegeticon* Oppian recommends several cunning ways to use this phenomenon for breeding horses and pigeons of specific colors. The famous Avicenna told the story of a hen that was sitting on her eggs when she was severely frightened by a falcon; the hen's fear penetrated the eggshells, causing the chickens to be born with falcon's heads.

One of the most famous instances of maternal impression occurred in thirteenth-century Rome. In the first year of Pope Martin IV's pontificate, a woman belonging to the noble family of Ursini gave birth to a child with the fur and claws of a bear. Men of learning attributed this birth defect to a painting of a bear which she had kept in her bedchamber. When this cause was explained to the pope, he ordered, somewhat irrationally, that all pictures and statues of bears in Rome were to be destroyed. Some theologians thought the pontiff had let the woman off too easily; they recommended that all bear wards in Rome be asked whether one of their beasts had broken loose nine months earlier and that the woman be tortured to extract a confession of her inordinate fondness for bears.

The Fish Boy and the Cat Woman

During the Renaissance period, the theory of maternal impression was extensively discussed, and philosophers and medical men agreed that it was possible. In his tract on Genesis, Martin Luther wrote that it should be considered one of the most certain principles in medicine. It was generally believed that if the pregnant woman longed for or was frightened by some object, and simultaneously touched some part of her body, the child would have a birthmark on the corresponding part, a kind of "psychic tattoo" of the object in question. The mother's soul was thought to be linked to that of her fetus, and the "animal heat" would bring the blood to the same part of the body in both of them, causing the mark on the tender flesh of the fetus. Frequently, the mother's desire for some particular type of food marked the child's skin, particularly if her longing were baulked. These birthmarks on the baby's skin were often compared to the form and shape of strawberries, cherries, or grapes. During the early 1600s, many scientists rejected the old notions that the birth of monstrous children was due either to divine displeasure or to the mother's copulation with a demon or an animal. The maternal impression hypothesis had a far greater appeal to the learned and the ignorant. Certainly, it was a mercy to the unfortunate mothers of malformed children, who had previously had to endure grueling interrogations and even torture from strict inquisitors who suspected them of alliance with Satan.

The year 1608 saw the first longer book devoted to maternal impressions, *De viribus imaginationibus tractatus*, written by Fienus. One of his examples of the action of the pregnant woman's imagination is also illustrated in an Italian handbill, which depicts a Neapolitan woman who is greatly frightened by some sea monsters and later gives birth to a son who is scaly all over like a fish. This fish boy of Naples was said to eat only fish, and to smell like one too. These attributes did not diminish his power of attraction for the hordes of curious spectators. Soon, he had become a European celebrity, touring France, Germany, Holland, and even Sweden with considerable profit. In reality, this fish boy must have suffered from severe ichthyosis, a congenital disease that gives the skin a scaly appearance.

Many sixteenth- and seventeenth-century medical scientists and biologists of great repute embraced the maternal impression creed. One of them was the famous French surgeon Ambroise Paré, who mentioned this theory as one of the nine causes of monstrous births. In his *Des monstres et prodiges* he produced the example of a pregnant lady who held a frog in her hand and later gave birth to a child with a frog's head. The anatomists Caspar Bauhin and Jean Riolan were other believers, as was the great Dutch anatomist and zoologist Jan Swammerdam. He gave a peculiar case report in his *Uteri mulieris fabrica*. A pregnant woman was frightened by a black man, but she realized the danger and hurried home to wash her entire body to prevent her unborn baby from turning black. When the baby was born it

148   was black in all the places she had not been able to reach while washing! Another Dutch anatomist, Nicolas Tulp, who was immortalized in one of Rembrandt's paintings, described a woman with such a passion for salt herring that she consumed more than fourteen hundred of them during pregnancy; the child had barely drawn breath before it, too, was heard to call out loudly for salt herring! In his *Experimental Philosophy*, Robert Boyle related the story of a woman who had gazed long and earnestly on some red pebbles near St. Winifred's Well; as a result, her child had red marks on its skin. It is fortunate that Boyle spent less time trying to prove silly medical theories than pursuing his work in experimental physics and chemistry, where he made the groundbreaking discovery of Boyle's Law relating the volume and pressure of gases.

Thomas Bartholin, the famous Danish anatomist, may have been the foremost of all seventeenth-century champions of the theory of maternal impression. His writings abound with bizarre case reports of women bearing malformed children after various untoward incidents during pregnancy.

Fig. 1. The fish boy of Naples; from an undated, late seventeenth-century Italian print in the author's collection.

When Bartholin visited Holland in 1638, he saw a girl with a cat's head, formed when a cat hiding in her mother's bed dashed out unexpectedly and startled the pregnant woman. One Copenhagen wife was frightened by an impudent beggar, who was both one-armed and lame; she later gave birth to a child with the same stigmata. Another of Bartholin's pregnant patients struck her head against a sack of coal; her child had black hair on half its head, white hair on the other. In 1673 an infant girl was born in Copenhagen with its hair already curled and frizzled according to the current style. Some strict clergymen took this portrait as an illustration of God's repugnance for the vanity and coquetry of the ladies of fashion, but others suspected that another maternal impression had taken place, after the child's mother bumped into one of the queen's ladies-in-waiting, in full court dress, while walking in the street. Thomas Bartholin, for once, was skeptical; the child's likeness to the lady-in-waiting was exaggerated, and the whole story had been distorted by superstitious people. Thomas Bartholin's son-in-law, Oliger Jacobaeus, told the story of a pregnant Dutch woman who had seen a dancing dog perform in the street; her child had deformed legs resembling the dog's. Thomas Bartholin and Oliger Jacobaeus were both close to the Danish royal family, and in view of the many dire maternal impressions occurring in his capital, King Frederik IV seconded the proposition that invalids and malformed people should be kept out of sight, in a special hospital in Copenhagen, not out of pity for the poor cripples, but to prevent pregnant women from bearing children exactly like them.

Maternal impressions were of common occurrence all over Europe during this time, and even the most drastic examples were freely accepted not only by the common people but also by scientists and medical men of repute. In the year 1595 a baby boy in Silesia was alleged to have been born with a gold tooth in its mouth. Soon, a hot debate had begun: was this a portent of divine displeasure or had the child's mother gazed too closely at some gold object? Several learned Latin pamphlets were written on this burning question, and the parents earned much money by exhibiting their child before the populace until a clever goldsmith solved the riddle by scraping off the gold covering the deceptive parents had put on their newborn infant's tooth.

When a noble French lady, Madeleine d'Auvermont, gave birth to a healthy child in 1637, her relatives were aghast, since her husband was well known to have spent the last four years abroad. She was brought to trial for adultery before a court of stern-faced priests and noblemen. Her defense was that she had thought intensely of her departed spouse and had dreamed about him at night; the child had been conceived by the power of her imagi-

150   nation. After medical and theological experts had been consulted, the verdict from the Parliament of Grenoble was that the miraculous infant was to be considered a legitimate son and heir to M. Hieronyme Auguste de Montleon, the husband of the clever Madeleine. Even Thomas Bartholin quoted this case, though his manner was rather tongue-in-cheek.

In the Jesuit priest Father Malebranche's *Recherche de la vérité*, published in 1679, several new theoretical arguments were presented, illustrated, as usual, with macabre case reports. A Parisian lady, who had gazed at a picture of Saint Pius during pregnancy, gave birth to an infant closely resembling

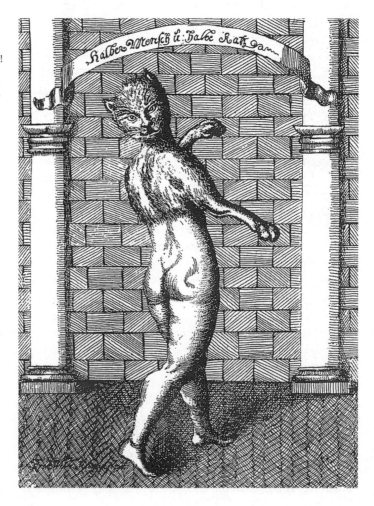

Fig. 2. Thomas Bartholin's alluring cat woman; even he had never seen a Batman! This illustration is from a 1679 German broadsheet in the author's collection.

the saint. The child apparently died soon after; the learned father remarks that he, like most of Paris, had seen it preserved in spirits. Another French lady had seen a criminal being tortured and broken on the wheel; she was later delivered of an idiot son with all his limbs fractured. Malebranche proposed that "all the blows given to the Malefactor did violently strike the Mother's Imagination, and, by a Counter-blow, the tender and soft Brain of the Child." The infant's brain was irreparably injured, and the mother's "animal spirits" had flowed toward the parts that had been injured on the malefactor's body. Her own tissues withstood this powerful flood of feeling, but the child's weak bones broke and crumbled.

The Debate after the Mary Toft Scandal

The strange story of the peasant woman Mary Toft of Godalming in Surrey, who had tricked King George I, the Prince of Wales, and England's leading men midwives into believing that she was giving birth to rabbits, added much fuel to the English debate on maternal impression. One of the many pamphlets inspired by this obstetrical farce was titled *The Strength of Imagination in Pregnant Women Examin'd*. The anonymous pamphlet writer ridiculed the Rabbit Breeder's assertion that her reproductive system had changed in a sinister way as an effect of her unfulfilled longing for rabbit meat. A certain Dr. Daniel Turner, who had written a book on skin diseases and who had proved himself a wholehearted supporter of the theory that a pregnant woman's imagination could cause monstrosities, was harshly criticized. Two years later, the nettled Dr. Turner replied in a pamphlet of his own, collecting many dubious cases from Bartholin, Fienus, and Malebranche. Later in 1729 his opponent disclosed himself to be Dr. James Augustus Blondel, a well-known London physician, in a book titled *The Power of the Mother's Imagination over the Foetus Examin'd*. This 143-page volume was the deadliest blow yet struck against the old fallacy, and unlike the works of Dr. Turner and other "imaginationists," it displayed a good deal of wit, intelligence, and common sense.

In the introduction Dr. Blondel states, "My design is to attack a vulgar Error, which has been prevailing for many Years, in Opposition to Experience, sound Reason, and Anatomy: I mean the common Opinion, that Marks and Deformities, which Children are born with, are the sad Effect of the Mother's irregular Fancy and Imagination." He reexamines all Turner's odd case reports and criticizes them soundly, directing biting sarcasm toward the hapless Dr. Turner and other, older proponents of the belief. The elder

Schenckius had described a woman uncommonly big with child, whose imagination had been caught by a religious parade. She subsequently gave birth to triplets in the likeness of the Three Kings; one of them resembled an Ethiopian. Blondel thought that "this story is very proper to be added to the Voyages of Captain Lemuel Gulliver, a Gentleman reported to be of such a veracity, that he was never catched in a Lie." Another of Turner's tales, fetched from the works of Fienus, concerned the wife of an actor in Brabant, who had had intercourse with her husband when he was dressed in his stage costume as the Devil; her child was duly equipped with horns, tail, and cloven hoofs. Blondel queried, "But pray, what should fright Jack-Puddings wife? Was she not used, and did she not delight to see her husband in that odd Dress?" Ambroise Paré's frog child, whom he derided as "Parey's Frog," also came in for criticism, for how could a frog scare "a Woman who was wont to make Fricassees of them instead of chickens?" The most horrific story of all was that of "Miss Muscle and the Grenadier." Canon Philippus Meurs, of Saint Peter's in Louvain, had told the chronicler Fienus that he had had a little sister with a great sea mussel between her shoulders instead of a head! This grotesque monstrosity was of course the sad effect of Mme Meurs's excessive desire for shellfish while she was pregnant. To feed "Miss Muscle," as Blondel called her, it was necessary to spoon hot soup into the gaping bivalve. One day, while she was being fed in this way, she bit the spoon angrily, broke the shells, and died soon afterward at the age of eleven. Blondel exclaimed, "Who ever heard the like? A Muscle fed with a Spoon! *Credat Judaeus Apello, non ego.*" He returned to the original source and found that not even Fienus had accepted this ludicrous fable at face value, since Canon Meurs was a very old man and often disposed to be untruthful in his utterances. Dr. Turner, however, did not doubt the elderly cleric's veracity. He added a case he himself had observed in London: a child born with a fleshy excrescence on its head resembling a grenadier's cap.

James Augustus Blondel introduced several new theoretical arguments against the old fallacy in his book. He noted that the proposed causes for various malformations were of much more common occurrence than the deformities themselves. Many women were frightened by evil-looking invalid beggars or had perilous encounters with frogs, mice, and hares without any ill effect to their children. Dr. Blondel had himself been consulted many times by nervous ladies worried about some strange craving or fright during pregnancy, but he had always calmed them down, and they had subsequently given birth to healthy children. He told of "a Gentlewoman, of very good Credit," who had been attacked by a large cat, which leapt into her bedchamber through the casement; the servants had to be called to pull the infuriated feline off her. Nevertheless, her son "had not the least Shew of *Smellers* or *Claws*," and far from being frightened by cats, he "loved to handle, and even to torment them, as if he took pleasure to revenge the Insult his Mother had suffered." Another lady blamed some reddish spots on the cheeks of her infant on the claret she drank during pregnancy. Following his jocose arguments in the book, Dr. Blondel might have recommended a rosé so as to achieve a healthy blush.

If the theory taken seriously by Dr. Turner and other worthies, that a pregnant lady might be impressed by the painting of a bear and have a child with bearlike characteristics, were to be universally accepted, this would have dramatic consequences for the future of zoological iconography. First and foremost, certain signs of London public houses, like the "Spread-eagle" and the "Hog in Armour" must be torn down, being immediate risk factors for engendering the most hideous monstrosities in the wombs of unsuspecting ladies passing by them. Blondel also noted the lack of a physiological model for the action of the imagination: "Let the Blood and Spirits be in never so great a Hurry, they can't do the Office of a Musket Ball, of a Hammer, or of a Knife."

Despite Turner's objections, Blondel was the undisputed victor in the debate, and there was a perceptible change in opinion not only in England but all over Europe, for Blondel's book was translated into French, Italian, Dutch, and German, and several other pamphlets followed in each country, often with much debate. In France Dr. Isaac Bellet published a lengthy pamphlet titled *Lettres sur le pouvoir de l'imagination des femmes enceintes* in 1745. Addressed to a female public, it dealt less with anatomy and physiology than with anecdotes and commonsense advice. Like Blondel, Bellet focused on Father Malebranche, who had been the foremost French proponent of the theory of maternal impression. In Germany the great weight of Herman Boerhaave's authority was in favor of mental impression as a cause for monstrosities, but Albrecht von Haller agreed with Blondel and added several important arguments. He pointed out that the birthmarks likened to strawberries and grapes were in reality small congenital skin tumors. Like Blondel, but with more experimental evidence, von Haller denied the existence of a direct nervous connection between mother and fetus. Without such a connection, he considered it impossible that complex messages from the mother's imagination could be conveyed to the child in utero. He also discredited the old notions that the signal transduction occurred by way of the blood, having no knowledge of hormones and cytokines!

154   Linnaeus and William Hunter on Maternal Impressions

During his famous journey to Lapland in 1732, Carl Linnaeus visited the small town of Piteå, where he was consulted by the parents of a child with a bad squint. Seeking the underlying cause of this phenomenon, he asked the child's mother "if she had seen someone squint or cock the eyes in a similar manner, while she was pregnant." The astounded woman, no doubt marveling at the skill of the famous doctor from Stockholm, answered that her mother-in-law had died during this time. "Hinc illis lacrymae," Linnaeus wrote laconically in his travel diary.

In 1756 the Imperial Academy of Sciences at St. Petersburg proposed to award a prize to the thesis that could explain why the body of the fetus was marked when the mother's mind was disturbed by some powerful stimulus. The prize was won by Dr. Karl Christian Krause of the University of Leipzig, who suggested that there was a nervous connection between mother and child through the umbilical cord. A more distinguished scientist, Professor Johann Georg Roederer, also entered the contest. His thesis was a concise summary of Blondel's and von Haller's arguments against the old fallacy. The formulation of the question, however, hints that the academy had made up its mind about the main question beforehand.

A few years later, in 1759, the Berlin Academy of Sciences discussed a peculiar effect of *das Versehen*. A tiny lapdog had strayed into an enclosure for birds, where she was violently attacked by a large turkey cock. Later, the dog gave birth to a malformed puppy, whose head resembled that of a turkey. A certain Dr. Eller, who presented the case, rejected the hypothesis of the Berlin townsmen that the lapdog's fright had affected its unborn puppy, but he favored an even wilder speculation. He proposed that the dog must have eaten one of the turkey's eggs, whose minute molecules melded with those of the puppy and altered its shape within the bitch's womb. This thesis gave the German proponents of maternal impression a golden opportunity for counterattack. They ironically explored the new prospects for dog breeding opened by their learned colleague, speculating on the wonderful things that might be achieved by feeding pregnant bitches eggs under scientific tuition.

In the 1790s the medical men of Berlin took up the debate again. A German woman, who was married to a black man, had given birth to a perfectly white child, and she tried to fend off the cuckolded Moor's rage by claiming that she had been greatly "impressed" by a painting of a white man which had hung in her apartment while she was pregnant. This time-honored last resort for unfaithful wives for once proved ineffective. The court submitted the case to the Medical Board of Berlin, and its verdict was that a black man could under no circumstances whatsoever be the father of a white child.

In England the debate continued throughout the late 1700s, the majority of men of learning taking Blondel's side. The anatomist William Hunter proposed an infallible plan for determining the truth of this age-old teratogenic theory; all women who entered one of London's largest lying-in hospitals were to be asked whether they had had any untoward experiences during pregnancy which they suspected had affected the unborn child. After birth, the attending men midwives were to see how many — or rather, how few — were right in their predictions.

In France, opinion was rather more in favor of the old doctrine, and the *Histoire des anomalies* by the great teratologist Isidore Geoffroy Saint-Hilaire contains an amusing story illustrating that not even a revolution could eradicate the maternal impression belief. In the third year of the French Republic, an infant was born with a birthmark resembling a Phrygian cap of liberty on its chest; its mother was awarded a pension of four hundred francs per annum by the government, probably for her patriotic thoughts!

Maternal Impressions among the Victorians

In the early 1800s the majority of medical men denied that a fetus might be disfigured through psychic imprinting, nor did any of them believe that an eight-month fetus might lose its arm when the mother saw a one-armed beggar. The notion that the mother's state of mind during pregnancy determined the child's appearance still lingered, however, and the "imaginationists" even regained some ground during these years. In Paris it was common practice among noble ladies to have themselves wheeled around the Louvre in easy chairs during pregnancy, with long stops for them to gaze at the paintings they wished their future children to resemble. Among the English common people, the old fallacy still reigned supreme, as evidenced by a curious broadsheet from 1817. A wicked adulterer had lied upon oath to avoid being named as the father of an illegitimate child, but the mother had thought so intensely on her seducer that his name, John Wood, was imprinted into the eyes of the child for all to see. These letters are likely to have been opacities in the irises, but there may be another explanation. Dr. Blondel had seen another child, with the words IEHOVAH ELOHIM in its eyes, exhibited in London in the 1720s. A committee of rabbis from the London synagogue

A FULL and PARTICULAR ACCOUNT of that most wonderful CHILD from Galloway, now exhibiting in this City, for the inspection of the curious; to which is added, a correct likeness, drawn from the life.

A YOUNG woman in Galloway having proved with child, laid the same to a respectable man of the name of John Woods, who denied being the father of the same, and persisted in his denial saying that he would never acknowledge the child, unless his name *was written at full-length* on its face; and he accordingly gave his solemn oath before the Court to that effect. This made so much impression on the mind of the young woman, who was present, that his name and person remained constantly in her mind's eye, and when the child was born, the name of the father appeared in legible letters in the child's eyes, the name of " JOHN WOODS," on the right eye, and " BORN 1817," on the left eye. When John Woods, the alleged father, came to know this circumstance, he instantly absconded and has not since been heard of.

This wonderful child has now arrived in this city, and has been inspected by the Professors, and other learned Faculties of this city, and pronounced to be a most wonderful phenomenon of nature, and an astonishing dispensation of Providence in pointing out the truth against the wicked and perjured ways of men.

An inspection of this child will, it is hoped, be a salutary warning to all young persons of both sexes, first to beware of all such doings, and second to beware of perjury in their attempts to conceal their shame.

Fig. 3. A peculiar maternal impression from early nineteenth-century Britain. From a handbill in the author's collection.

went to examine this wonderful child and accepted him as the Messiah, but a clever oculist revealed that the boy's parents had fitted him with "two thin Pieces of painted Glass, commonly known by the name of Artificial Eyes"!

The theory of maternal impression was finally disproved during the mid-1800s, and after the year 1860, it received little support from serious scientists. By then, the embryological discoveries of the great German physiologist Johannes Müller and the anatomists Theodor Bischoff and August Förster were able to extend Blondel's arguments against the old superstition considerably. By this time, it was well known that many malformations were congenital and wholly independent of the mother's imagination. It had come to seem absurd that a seven- or eight-month fetus would, for example, lose its arm when the mother was frightened by a one-armed man. During the first weeks of pregnancy, when the fetus was most sensitive to disturbances in its development, the mother was not aware of its existence. Moreover, twins often had birthmarks at different sites, and the malformations traditionally associated with various animals were also occurring in countries where these animals did not exist. Another contradiction in terms was that the woman could not be "impressed" by herself; if she broke her leg or got a bruise, the child should have been marked as easily as if she had seen a hare leaping by her in the fields. It was even possible to deliver a perfectly healthy child through cesarian section from a dead mother — hardly feasible had her state of mind determined its mental and physical condition. During the late nineteenth century, the German teratologists August Förster and Gustav Schwalbe built up a system of malformations, which could all be related to some disturbance in the development of the fetus. This limited list disagreed completely with the belief of the "imaginationists" that the number of malformations was limitless, decided only by the range of the mother's imaging power. For example, it was by now understood that the monstrous infants described as having a cat or frog's head all belonged to the type acephalus. Toward the end of the century, the embryologists also had some understanding of the anatomical background of malformations such as conjoined twinning, anencephaly, and cleft palate.

Long after it had become obsolete among the medical profession in Germany and France, maternal impression still had many believers in Britain, not only old-fashioned country practitioners but also several hospital consultants of repute. There was a steady sprinkling of case reports in the *Lancet* and the *British Medical Journal* well into the 1890s. In 1890 and 1892 the distinguished Scottish obstetrician Dr. J. W. Ballantyne reviewed the whole subject in two lectures before the Edinburgh Obstetrical Society. Some modern cases had obviously impressed him greatly, and despite the growing mass of evidence against the old belief, he had to declare the matter *sub judice*.

In the United States, the old fallacy also died hard. According to the Index Catalogues of the Surgeon General's Office, more than 170 articles on maternal impression were published in American scientific journals between 1839 and 1920. Even respected obstetricians such as Dr. Fordyce Barker accepted this teratogenic theory. In 1870 the teratologist Dr. George J. Fisher of Sing Sing, New York, who was well read in the current European literature on the maternal mental influence, tried to call his country-men to order in a long article in the *American Journal of Insanity*. He was appalled at the widespread acceptance of the maternal impression theory, not only among American ladies of all ages, but also among his own colleagues. Dr. Fisher had himself many times encountered nervous pregnant ladies who had had some unpleasant experience, but none of them had any cause for their apprehensions. He concisely summarized the modern arguments against the old belief and insisted that "it is the duty of the medical profession to endeavor to correct this popular error by teaching the people the absurdity of their 'unquestioning faith.'" His arguments were to little avail. In 1894 Dr. Marcus P. Hatfield published a similar article in the *Transactions of the Illinois State Medical Society* of 1894. It brought no credit on American medical science, he declared, that the sponsors of the maternal impression theory were almost entirely confined to the United States. He was answered in no uncertain terms by Dr. Frank A. Stahl of Chicago in the *American Journal of Obstetrics* two years later. This gentleman had seen a child born with acrania (absence of the skull) after its mother had seen another child having its head crushed by a streetcar, and this experience carried more weight with him than all the theoretical arguments against the old fallacy combined. The American fascination with maternal impressions continued well into the 1900s. In the 1903 edition of the *American Textbook of Obstetrics*, Professor Edward P. Davis of Philadelphia boldly stated that the association of birth defects with these prenatal influences was more than mere coincidence: the famous Elephant Man of England and the Turtle Man currently on exhibition in the United States were hideous examples of the reality of maternal impression. Several case reports, written according to the time-honored method of Thomas Bartholin and Father Malebranche, were published in American medical journals shortly before the First World War. It is interesting to note that the fair sex made their first contributions to the literature on maternal impressions about this time: two of the reports were written by Drs. Belle Gurney and Jennie Gray.

In popular books on the workings of the body and in old-fashioned matrimonial guides for newly married women, the old fallacy survived even longer. The ever-popular *Aristotle's Master-piece*, first published in 1684 and containing a wealth of misinformation about maternal impressions and many other subjects, was reprinted without interruption well into the 1930s both in Britain and in the United States. Some of the firmest early twentieth-century supporters of the maternal impression theory could be found in the early Christian Science movement.

A peculiar offspring of the maternal impression creed was telegony, the belief that influences might recur from earlier conceptions; thus, the child of a remarried woman might resemble her previous husband. Philosophers such as Schopenhauer and Herbert Spencer considered telegony possible in both man and beast, and Charles Darwin discussed it in his *Variation of Animals and Plants under Domestication*. Darwin was particularly impressed by a case originally reported by Sir Everard Home, a distinguished surgeon of the early nineteenth century, who accepted all aspects of maternal impression. An Arabian mare belonging to Lord Morton had been mated with a quagga, a zebralike wild donkey of Africa, which is now, alas, extinct. After producing one hybrid foal, the mare was mated with an Arabian stallion, and to his lordship's great consternation, this foal was also striped like a quagga. Portraits of all the animals involved are still kept at the Royal College of Surgeons Museum in London. In the late 1800s the idea of telegony was further advanced by several racial theorists: a woman who had once had a child with a man belonging to an inferior race would never, it was said, be able to bear a child of pure Aryan race again.

In the 1890s Weismann and Ewert laid the foundation of modern genetics, and telegony was disproved. Some conservative livestock breeders remained firm adherents of telegony, believing that once their cows had mated with a prize bull, the bull's desirable characteristics would be conveyed to subsequent progeny, even though an inferior bull was employed thereafter. Telegony has also been a pet theory of twentieth-century racists and anti-Semites. Some Nazi propagandists held that a German maiden who had once defiled herself with a Negro or a Jew, could, to her eternal shame, never give birth to a child of pure blood again.

Maternal Impressions in Folklore

A century ago, pregnancy was a time of great peril for the expectant mother. The old wives had a complex system of regulations and taboos to prevent maternal impressions from exerting their sinister effects. If these regulations went unheeded, the child was sure to be "marked" in some way. This belief was current over large parts of the globe, and the examples and regulations are very similar in Scandinavian, German, English, and American folklore, speaking in favor of very ancient origins for these notions.

A disorder often attributed to maternal impression was epilepsy, which might be caused by the sight of a slaughtered animal falling to the ground or by a fall suffered by the mother herself. If the expectant mother walked over a water-filled ditch or made water in a churchyard, her child would be a bed wetter; if she squinted through a keyhole, the child would be marked with a squint for its whole life; if she climbed over the shafts of a carriage, her baby would be bandy-legged; if she ate speckled bird eggs, the child would be heavily freckled; if she helped to shroud a corpse, the child would be pale and sickly; and if she reached for some object or climbed under a rope, the umbilical cord might be wrapped around the baby's head. An alarming notion, current in both Germany and Scandinavia during the nineteenth century, was that if the expectant mother carried logs of wood in her apron, the (male) child's genitals might attain an enormous size. A particularly widespread belief was that if the mother, frightened by some object, grasped some part of her body, the child would have a birthmark depicting the object in question on that body part. Thus, the mother was to slap her back after being frightened to keep from disfiguring her unborn child too much or, even better, to strike the ground to discharge the energy of psychic imprinting.

The old notion that the mother's longing for some particular type of food might mark the child's skin was also very common in folklore. The child's birthmarks were likened to strawberries, plums, and black currants. This superstition seems to have changed with people's eating habits. Modern American folklore from Utah mentions birthmarks in the shape of a slice of bacon or even a roasted broiler! We have no record, nevertheless, of upperclass ladies having children marked with *grenouilles* or *paté de foie gras*.

The expectant mother had to be particularly careful in her dealings with members of the animal kingdom. If she saw a duck, the child might be born with webbed hands and feet; if she was frightened by a snake, the baby might have staring eyes and a flickering tongue; if she kicked a pig, the child might speak in grunts through its nose. The notion was very widespread that the child's body would be marked with the shape of a mouse, fur and all, if its mother was frightened by one of these creatures. The idea can probably be traced to the occurrence of the so-called benign hairy nevus, a rather large congenital skin tumor, which might well be the size of a mouse. The most dangerous animals of all were the rabbit and the hare, which could mark the unborn child with a harelip. A Swedish collection of folklore contains no fewer than 121 instances of this superstition, and it occurs frequently in German and English folklore too. Its geographical distribution seems to equal that of the hare. In the *Historia de gentibus septentrionalibus* by Archbishop Olaus Magnus, published in 1555, this learned writer mentions "an unfortunate accident, often happening to pregnant women, is that they by eating hare-meat, or treading on a hare's head, bring forth children marked with a hare-lip, whose lip is always cloven between the nose and the mouth. The only cure is that you take a small piece of meat, which should still be bloody, from the chest of a tiny chicken, which has just been slaughtered, and stitch it to the part in question." The tissue transplantation suggested by the imaginative prelate is not feasible even with modern methods of controlling the immune system in order to prevent rejection.

Not only the child's outward appearance but also its character and behavior might already be irreversibly decided in the womb. Scandinavian and German folklore agreed that the expectant mother who stole some petty object would bear a child who would become a thief; accepting some object over a doorstep or climbing in through a window would have the same effect. If she spilled beer on her clothing, the child would become a drunkard, and if she swore and cursed, the child would be born insane. If she looked down into an open grave, the child would have a voracious appetite and sit with its

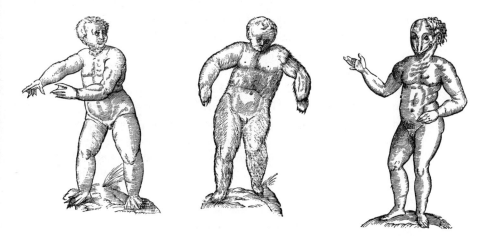

Fig. 4. Frog boy, bear boy, and duck boy: three frightful examples of maternal impressions from Ulysses Aldrovandi's *Monstrorum historia* (1642). From the author's collection.

jaws wide open all day, screaming for more food. American folklore gathered in Utah during the mid-1900s includes several similar examples, although less drastic; for instance, a woman could make her child a talented musician by playing the piano a good deal during pregnancy. A particularly widespread notion was that a woman should not think of men other than her husband during pregnancy for fear of engendering a likeness that might disturb the married couple's future happiness.

Fortunately, there were popular cures for many of these malformations. Most followed the time-honored principle of seeking the cure where you got the ill. If the child had been struck with the falling sickness after the mother had witnessed an animal being slaughtered, it could be cured by drinking blood from a newly killed animal. A birthmark caused by the mother's being hit by hot grease from a frying pan could be cured by rubbing with greasy bacon. In the same fashion, a "fire-mark" or hemangioma could be washed away with ashes from a freshly burned-out fire.

Maternal Impressions in Literature

Since the sixteenth century, many writers have alluded to the maternal impression theme in their works. Philosophers have used this motif for enlightening moral debate. Was the character already decided in the womb, for example, and would such predetermination still allow for the dogma of God as the ultimate Creator? Popular novelists often used the motif of the woman who received some shock during pregnancy and later gave birth to a malformed child to advance the plot or heighten dramatic tension, and to refer to a theory well known among their readers.

Already in the sixteenth century the maternal impression theme had a prominent part in several major poetical works. The first of these was the Italian Torquato Tasso's epic *Jerusalem Delivered*, originally published in 1581. Tasso reused the plot of Heliodorus's *Ethiopica*; the white-skinned heroine Clorinda is the daughter of the Ethiopian King Senap and his queen. During pregnancy, she had gazed on a statue of the virgin in the legend of Saint George and the dragon, and the child's features exactly matched those of the statue:

> *At length her womb disburthen'd gave to view*
> *(Her offspring thou) a child of snowy hue.*
> *Struck with th' unusual birth, with looks amaz'd*
> *As on some strange portent, the matron gaz'd:*

> *She knew what fears possess'd her husband's mind,*
> *And hence to hide thee from his sight design'd.*

William Shakespeare frequently hints at maternal impressions in his plays, never more explicitly than in the third part of *Henry VI.* When Queen Elizabeth hears the bitter news that King Edward has been taken prisoner, she tries to command herself, since she knows that her violent passions may harm her unborn son in the womb:

> *Ay, ay, for this I draw in many a tear,*
> *And stop the rising of blood-sucking sighs,*
> *Lest with my sighs or tears I blast or drown*
> *King Edward's fruit, true heir to the English crown.*

In 1709 the French abbot Claude Quillet published the long poem *Callipaedie,* a disquisition addressed to mothers on how to bear healthy and beautiful children. Maternal impressions were of course not neglected by the learned abbot, who discoursed at length on their mechanism. He believed that tiny "atoms" were formed each time the eye beheld some object and that their nature depended on the character of the objects that met the eye. These sinister atoms could travel far within the body:

> *They gain the warm Recesses of the Heart,*
> *And are from thence diffus'd to ev'ry Part,*
> *From thence they to the Womb their Passage make,*
> *And young Conceptions thus their Likeness take.*
> *Fair if the Object, will the Atoms be,*
> *And with their Shape the future Birth agree.*
> *From a foul Figure if the Image flows*
> *The* Foetus *foully like the Object grows.*
> *The Soule and Eyes it will at once offend,*
> *And filthy Atoms on the Womb descend.*

In 1718 the first English edition of the *Callipaedie* was published, together with a translation, by Tytler, of an older Latin poem titled *Paedotrophia,* which was written in the late 1500s by a certain Scevole de St. Marthe. In this poem the actions of the pregnant woman's imagination are fully explained:

> *Then (wonderful to tell!) if you deny*
> *The strange request, nor with their wish comply,*
> *Avenging Nature, from unknown designs,*

With spots and marks the foetus' body signs
With stains indelible, that never can
Wear out, thro' life, in woman, or in man.
And — stranger still! — while in the mother's breast,
This passion sways, and rages o'er the rest,
Whatever place she scratches, or besmears,
A mark, in the same part, her infant bears.
Hence oft unseemly moles and freckles grow
On virgin bosoms white, besides, as snow;
O'er beauteous bodies wens and tumours steal,
And for the mother's guilt, the daughters feel.

The first mention of maternal impression in English prose literature occurs in Mrs. Aphra Behn's novel *The Dumb Virgin; or, The Force of the Imagination*. In this sad story, a poor woman gives birth to a deaf-mute daughter and several other malformed children — all because of the sinister effects of her imagination. In a better-known novel, Henry Fielding's *Joseph Andrews*, Joseph has a strawberry birthmark on his chest as a result of his mother's longing for strawberries while he was in the womb; by means of this distinctive mark, he is recognized by his rightful mother in the novel's happy ending. In Fielding's novel *Jonathan Wild*, the character of this notorious criminal has been determined by prenatal influence; his pregnant mother has violent desires to acquire all sorts of property and, so, marks her unborn son for a thief. Another well-known eighteenth-century novelist, Tobias Smollett, was also a medical doctor and a supporter, if not a friend, of Dr. Blondel. In his novel *The Adventures of Peregrine Pickle*, which was published in 1751, Smollett ridiculed the excessive precautions taken by his countrywomen to avoid the "marking" of their unborn children. He mocks in particular the notion that no longing for any particular kind of food must be unfulfilled lest the child be born with a birthmark resembling the edibles in question. In the novel the pregnant Mrs. Pickle is guarded by her sister, Mrs. Grizzle, who holds to the old doctrine; poor Mrs. Pickle's every action is scrutinized by her stern guardian, and when Mrs. Pickle longs to eat a pineapple, Mrs. Grizzle tries every means to provide one so as to prevent the marking of Peregrine's skin with the image of this fruit.

During the eighteenth century, not only maternal but also paternal impression was considered quite possible; the father's state of mind during coition, when the child was conceived, was thought to determine its future character. Particularly widespread was the idea that coition during the menses produced only puny and deformed offspring — and redheaded at that! Similarly, if the father was drunk when the child was conceived, it was doomed to become an idiot. Laurence Sterne poked fun at these odd notions in his famous novel *The Life and Opinions of Tristram Shandy*. Just at the moment of Tristram's conception, Mrs. Shandy asks her husband if he has remembered to wind the clock, thus jolting young Tristram's dormant mind in a sinister way. Mr. Shandy laments this unfortunate mishap in long monologues, concluding that "my Tristram's misfortunes began nine months before ever he came into the world." Sterne also wanted to ridicule some wild contemporary embryological theories, which said that the human being was fully formed at conception, differing from the newborn infant only in its extremely diminutive size. In the book's amusing preface, this "little Gentleman" is considered to be "as much and as truly our fellow creature as my Lord Chancellor of England."

An even more ridiculous fad was lactational heredity — that the quality of the wet nurse's milk could influence the character of the child. A Victorian wet nurse had to present references, preferably from a clergyman, stating that she was of a good character, before she was employed by a cautious parent. Captain Marryat poked fun at these notions in his novel *Mr. Midshipman Easy*. When a wet nurse is to be employed for young Jack Easy, his father is apprehensive lest his son "imbibe with his milk the very worst passions of human nature." Only a prolonged phrenological examination of the wet nurse convinces him of her good character. Dr. Maurice Kenealy, member of Parliament, the somewhat unbalanced legal counsel of the notorious Tichborne Claimant, an enormously fat butcher who returned from Australia to lay claim to one of England's richest baronetcies, took lactational heredity more seriously. He once ascribed the immoral life led by the Prince of Wales to the sinister effects of "the foul, debased milk" of an alcoholic wet nurse employed by Queen Victoria!

Sir Walter Scott was another supporter of the theory, at least insofar as it might lend interest to the subplot of a novel. In *Redgauntlet* the heroine has "blood-stains" on one arm, received in utero through psychic imprinting when her mother witnessed her father's execution as a Jacobite rebel. In *The Fortunes of Nigel*, Scott refers to a particularly well known historical maternal impression: that affecting Kind James I; the spineless timidity of this monarch, as well as his abhorrence of swords and daggers, was decided when Darnley and his men stabbed Rizzio to death before the eyes of the horrified Mary Queen of Scots.

Charles Dickens provides another well-known literary maternal impression in *Barnaby Rudge*. Barnaby's father, a robber and murderer, comes home late one night splattered with blood and frightens his pregnant wife. A drop of blood falls on her wrist and the unborn Barnaby is marked at the same site. Furthermore, Barnaby's excitement when gold and riches are spoken of, his feeblemindedness, and his great fear of blood are all said to have been established long before he was born, together with his disposition toward desperate and dangerous adventures, which leads to his imprisonment as a participant in Lord George Gordon's "No Popery" riots.

In his article "L'imagination" in the *Dictionnaire philosophique*, François Voltaire accepts the idea that the imagination sometimes marks the unborn child according to the mother's impressions. A passage in *Träume einer Geisterseher* hints that Immanuel Kant also accepted the old fallacy. Johann Wolfgang von Goethe's famous *Die Wahlverwandtschaften*, published in 1809, contains a world-famous combination of maternal and paternal impression. When a married couple makes love, each partner thinks of his or her unfulfilled passion for another person, and the resultant child resembles these others. Goethe's contemporary Jean Paul Richter made burlesque use of the old fallacy in his novel *Dr. Katzenberger's Badereise*. The unsavory doctor, the owner of a chamber of horrors containing several monstrosities pickled in alcohol, shows them to his pregnant wife in the hope that she will bear him a splendid new specimen for his collection.

The theory of telegony also appears in several literary works of repute. Emile Zola mentions it in *Le docteur Pascal*, as does Henrik Ibsen in *The Lady from the Sea*. August Strindberg, whose fertile mind was always easily caught up by various scientific fads, was even more fascinated by these obscure speculations, which recur in many of his books and plays. In *Creditors* (1890), for instance, an artist named Adolf meets a schoolmaster at a spa, without knowing that this gentleman is the divorced husband of his wife. In their conversation Adolf mentions that he and his wife had to put their first child up for adoption because when it was three years old, it started to resemble the wife's previous husband, leading to several embarrassing situations. The clever schoolmaster answers that the previous husband should not be suspected of foul play, for it was well known "that the children of a remarried widow resemble her late husband. This is a pity, of course, but it was the reason for burning widows at the stake in India!"

Some literary historians have presumed that Strindberg was influenced by Goethe's *Wahlverwandtschaften*, but his knowledge of telegony went further than Goethe's. In the second act of Strindberg's play *The Father*, the insane

Captain asks Dr. Östermark if it is true that once a mare is crossed with a zebra, all its future foals, even those sired by a stallion, will be striped, This remark would imply that Strindberg knew the story of Lord Morton's mare and its amorous adventure with the quagga. It is very unlikely that he had studied the original source, Sir Everard Home's *Lectures on Comparative Anatomy,* but the story of the mare was repeated in several popular works. He might have encountered it in Charles Darwin's *Variation of Animals and Plants under Domestication,* a book that was well within the purview of Strindberg, who read widely, although erratically, in the natural sciences. Another odd notion from Darwin's book also recurs in Strindberg's work: that a woman herself might absorb some of her partner's physical characteristics through the semen. This idea seems to have worried Strindberg a good deal. It pained him that he had to give away parts of his genius to his wives and other amours, and during his period of insanity, he imagined that, since a part of himself was living within their bodies, he could share the sexual experiences of his former wives. Strindberg's fascination with telegony may well be related to a drama in his own life: the tempestuous and short-lived marriage to Siri von Essen, who had borne two children to her former husband, the army officer Baron Wrangel. Strindberg was a particularly jealous and possessive character, and this ménage caused him much pain, as evidenced by his semiautobiographical *Plaidoyer d'un fou.*

꙳ The belief in maternal impression is well-nigh unique in its wide distribution, in time and place, and its extensive popular support. In his famous *Pseudodoxia epidemica* of 1640, Sir Thomas Browne comments on the vulgar error that crystal is really the concretion of snow and ice: "Few opinions there are which have found so many friends, or been so popularly received, through all professions and ages." Sir Thomas Browne might have said the same thing about maternal impressions, a subject unaccountably absent from his book; it is unknown whether he, too, was an "imaginationist" or if he simply considered this notion too silly for words.

In the sixteenth- and seventeenth-century medical world, the maternal impression hypothesis found universal favor, for it was universally applicable, easy to understand, and without ties to the contemporary religious and philosophical issues. Furthermore, the seventeenth-century books and theses on this question, which were of course all written by men, often emphasize the spiritual inferiority of womankind and the danger that a woman's perverted cravings and emotions during pregnancy could wantonly alter the shape of the fetus conceived in perfection by the man. The many strange tales, ranging from the marvelous to the ludicrous, put forth as evidence tended to follow a *post hoc ergo propter hoc* reasoning. This means that if there was a temporal relationship between two events, such as, for example, the woman's being frightened and her giving birth to a malformed child, this implies a causal relationship between them.

Throughout the 1800s, an increasing body of scientific argument was amassed against the long-lived old fallacy. By the 1860s it was clear to physiologists and embryologists that the idea utterly lacked foundation in fact. The notion had a wholly negative influence on the science of teratology, for instead of systematizing the observations of various malformations, the medical men focused on the likeness between the deformity and the assumed mental impression in each separate case. In England and particularly in the United States many country practitioners seem to have been most reluctant to discard the old fallacy. It is possible that they were influenced by its extensive popular support. Well into our own times, ethnologists have traced remnants of the belief in the United States and in various parts of Europe. In a study of current popular beliefs concerning facial deformities, performed by the English dentist Dr. W. C. Shaw in 1981, more than 20 percent of those interviewed believed that a "portwine stain" was due to the mother's craving for strawberries or red cabbage, and not a few still considered the hare responsible for a cleft palate, one respondent even blaming the mother's failure to tear her dress to neutralize the impression before it hurt the child.

In the 1930s the psychiatrist Dr. Bernard Stokvis studied a large aggregation of Dutch women who believed their children to be victims of maternal impressions. Stokvis wanted to explain the factors underlying this delusion. In various psychological tests, these women showed hysterical traits as well as what Stokvis called a "magic mentality." These factors made them liable to hysterical rejection of an imperfect child, out of neurotic consciousness of guilt. This investigation would have merited publication merely for its demonstration of the still extensive popular support for the old fallacy, but unfortunately, it appeared only in Dutch, and sank into undeserved obscurity. Dr. Stokvis may well have been right that in the 1900s, women of neurotic traits of character were particularly liable to invoke the maternal impression explanation after having given birth to a child with some "mark," but in past centuries, the old fallacy had an almost universal popular support. It is natural for any individual struck by a personal tragedy, such as giving birth to a malformed infant, to seek some reasonable cause, some idea of why it happened to them and not someone else; this could not be given by the church or by the medical establishment. The maternal impression belief achieved both in a comprehensible and definitive way, since the manifold alleged causes for various deformities were of much more common occurrence than the deformities themselves; it was thus likely that the mother could think of some "impression" likely to have affected her.

Tailed People

Of a standing fact, sir, there ought to be no controversy; if there are men with tails, catch a homo caudatus!

Samuel Johnson, in conversation with
Lord Monboddo

IN THE OPENING CHAPTER OF GABRIEL GAR-
cía Márquez's famous novel *One Hundred Years of Solitude*, the strange life
story of a distant member of the adventurous Buendia family is briefly
related. He had a cartilaginous tail shaped like a corkscrew with a tuft of hair
at the end. After keeping this "pig's tail" hidden in a pair of baggy trousers
during forty-two years of complete celibacy, he met a butcher who offered
to chop it off with a cleaver. Unfortunately, the butcher bungled the amputation and the ancestral relative bled to death. The last of the accursed Buendia family, the infant child eaten alive by white ants, according to the proph-
ecy of the sinister Gypsy Melquiades, is also born with a tail.

Many readers must have marveled at the fertile imagination of the famous Nobel Prize winner, but Gabriel García Márquez's description of this
tailed man may well have been the result of factual observation. More than
160 cases of human tails are more or less carefully described in the medical
literature; several of the tails have been likened to pig's tails, and their owners

have sometimes been able both to wag their tails and to curl them. What is harder to believe in García Márquez's story is that the unfortunate Latin American died after the tail amputation. These caudal appendages do not contain any large blood vessels, and even a butcher should have been able to stanch the blood flow from the wound with a heated poker or a dab of tar. Although there was a paper in the *Australian Medical Journal* in 1884 with the eloquent title "Congenital Caudal Growth; Removal; Death," the human tail in this case was of quite other dimensions than the one in García Márquez's story.

❧ One of the longest-lived of the old teratological misconceptions was that particular races of tailed human beings existed in faraway places. Quite extensive areas on old maps of the world are marked with the words "homini caudati hic": humans with tails live here. Several accounts of tailed races survive from antiquity. Both Pliny and Ptolemy described tailed people in India, and Pausanias described *satyrids* living on islands in the Atlantic Ocean, redhaired and repulsive-looking beings with long tails. Marco Polo described tailed human races in Asia, and the most popular book of the fourteenth and fifteenth centuries, next to the Bible, the spurious *Travels* of Sir

Fig. 1. Tailed people described by Sir John Mandeville. From the author's collection.

172   John Mandeville, also mentions an encounter with a tailed tribe, which was probably lifted from Pliny's *Natural History*.

During the 1500s and 1600s, intrepid travelers seem to have been as little surprised to observe savage tailed races as the charter tourists of our own time are to see the leaning tower of Pisa or some similar attraction. At regular intervals, stories of tailed races were related by seamen and explorers. Some of these savages were said to have long hairy tails, and others had to make holes in the thwarts of their canoes, so that they could sit down. It was always added that the description was only preliminary, inasmuch as the wild people could not be approached with safety.

Some travelers located the tailed huumans in the Polynesian Islands; others believed them to live in India or China. Some writers claimed that Constantine the Great's innumerable bastards kept the predisposition for tail formation alive in the highlands of Turkey; others that sinister sects of heretics in the Pyrenees had been equipped with tails as a Divine punishment. The English chronicler Samuel Purchas had much to tell about tailed people in his *Pilgrimage*. The Kingdom of Lambri, for instance, "hath in it some men with tayles, like dogges, a spanne long," and the Sumatran islands were inhabited by "certaine people there called Daraqui Dara, which have tayls like to sheepe." Even William Harvey, the discoverer of the circulation of the blood, was convinced by the works of certain Dutch geographers that a race of tailed people inhabited the island of Borneo.

The social status of these tailed tribes varied greatly. In Turkestan the tailed people were scorned as "stinking tailed vermin." By stark contrast the regal family of Saurashtra in India, the so-called long-tailed ranas, was held to be predestined for greatness by their long caudal appendages, taken as evidence that they were descended from the tailed monkey god Hanuman. As late as 1806 Captain Samuel Turner in his *Embassy to Tibet* noted the claim of an Indian worthy that there existed "a species of human beings with short straight tails, which, according to report, were extremely inconvenient to them, as they had to dig holes in the ground before they could attempt to sit down."

During the early 1800s, the people most often accused of having tailed "missing links" among them were black Africans. Many imaginative explorers and missionaries speculated about the wild and ferocious nature of these tailed savages. The English traveler Frederick Horniman was more cautious in his *Travels from Cairo to Mourzouk*. He heard no accounts of tailed people during his long journey, "except from one person (but not a testis fide dignissimus) who placed them ten days south of Kano; he called them Yem

Yem, and said that they were cannibals." For their part, the Africans had similar suspicions about the uninvited white visitors, whose long trousers they viewed with surprise and distrust.

The best known of the alleged tailed black races were the Niam-Niams, who were supposed to live in the least explored parts of Central Africa. In the 1840s and 1850s several French anthropologists published articles about the Niam-Niams; one of them was the famous François de Castelnau. The Niam-Niams, it was reported, lived in a thick jungle near Kano, in a part of the country abounding in wild beasts. Other tribesmen attacked and killed these tailed people on sight. M. de Castelnau managed to interview eight men who had pursued a party of Niam-Niams through the jungle; they readily agreed that their foes all had soft, motionless tails, ten to fifteen inches in length.

Castelnau's report aroused intense curiosity among the European anthropologists, and several research expeditions were sent out in search of the Niam-Niams. They were less fortunate in finding these tailed people, however, and based their reports on hearsay alone. The Africans were apparently not unwilling to tell tall tales about the savage Niam-Niams, and the stories got wilder and wilder. Consider, for example, an article published by a certain Dr. Le Bret in a French scholarly journal. Although he had never seen any of the Niam-Niams or had spoken to any individual who had, Le Bret did not doubt their existence. Not even the most brutal Arabian slave traders were able to subdue these tailed cannibals; the slavers had to be content with filing their teeth in order to quench their desire for human flesh, or at least their capacity to chew it. In order to sell the tailed savages as full-price goods, the Arabs cut their tails off, but slave buyers and wholesalers soon learned to look for the tell-tale scar. The trade in Niam-Niams had almost totally ceased, Le Bret reported, because "the children of this race are of a particularly savage disposition, and even devour the children of their masters; of this, there are only too many melancholy examples."

The story of the tailed Africans was good newspaper material, and it attracted some curiosity also in London scientific circles. The anatomical showman Dr. Joseph Kahn had the attention of the whole metropolis when he advertised that a family of tailed Negroes would soon be exhibited at his notorious museum in Tichborne Street. At the well-attended premiere, the main attraction proved to be some wax statues of black natives in full gear, to which the clever doctor had attached long tails at the rear. Although Dr. Kahn's lecture was held in good order, according to the newspaper accounts, it is really hard to believe that the rowdy audience, furious at being fooled, would not have disrupted the devious doctor's oration with boos, cat-calls, and threats of bodily violence.

 ⮞ One of the most remarkable traveler's stories about savage tailed human races was told by the Swedish sea captain Nils Matson Koeping, who made a long sea journey to "Asia, Africa and many other heathen countries" in 1647. Koeping obviously rather fancied himself a naturalist, having a keen eye for curious animals and plants throughout his voyage. He saw snakes with two heads, "crawling with one end first one month, and the other end first the other month," and a woman who gave birth to a hairy child fathered by a baboon: "As soon as it had been born, it jumped onto a pole, then sprang up on a door, and later climbed up a high tree." He also saw a chameleon that subsisted on air alone, the soaring birds of paradise which lacked legs to stand on, and the "elephant-master" with a horn on its snout and armored plate on its back. In spite of all these marvels of the East, the most desperate and dangerous new acquaintances made by this intrepid Scandinavian explorer were the natives of the island of Nicobar. He described them as "an ugly people of stunted growth, blackish yellow in colour, with *tails* at their backs like cat's tails, which they whisked from one side to the other, like cats." The tailed natives greatly desired scrap metal for some reason, and they eagerly searched the ship for nails and pieces of iron. They wanted to trade gray parrots for iron, but the sailors refused; in a fit of fury, the tailed men wrung the birds' necks and ate them raw, on the spot. This encounter, which did not end until the sailors fired some cannon to scare the tailed natives away, was not Captain Koeping's only experience of their avarice and voracious appetites. A foraging party, consisting of one of the ship's mates and some sailors, was later sent to the island, but it failed to return. When the ship's company saw fires being lit in the forest, the captain rightly considered this a highly ominous turn of events. When the sailors went ashore in force, they saw the ship's longboat with all tack drawn, and, near a burned-out fire, the bones and other remnants of the tailed cannibals' hearty meal of human flesh.

 Koeping's dramatic tales were widely quoted by European geographers and anecdotalists of Oriental manners. They also reached the Scottish judge and scientific amateur Lord Monboddo, who believed that language was not unique to humankind and that the orangutan was closely related to human beings. In order to prove this novel concept, he wanted to establish that people could have tails, indeed, that the tailed state was natural to human beings and that many infants were born with tails not only in Asia and Africa but in the very heart of Scotland. To prove his hypothesis, he even wanted to burst into the chambers of women in labor, in order to examine the infants before their caudal appendages had time to atrophy. Lord Monboddo's copious monograph *Of the Origin and Progress of Language* has been favorably judged by modern historians of science, who have appreciated his role as a forerunner of evolutionary thought, but it brought him only ridicule from his contemporaries. He was unfortunate enough to clash with the formidable Samuel Johnson, who delighted in mocking the eccentric judge's notions. Boswell's *Life of Johnson* contains many severe judgments on Lord Monboddo's theories. Once, Dr. Johnson said that "other people have strange notions, but they conceal them. If they have tails, they hide them; but Monboddo is as jealous of his tail as a squirrel." His legal colleagues also teased Lord Monboddo. One of them, the famous Lord Kames, having asked Monboddo to precede him into a room, remarked, "Just to see your tail, my Lord." Of the wits and parodists of his age who laughed at him, Thomas Love Peacock had the greatest success with his novel *Melincourt*, the hero of which is an orangutan called Sir Oran Haut-ton, which is captured while young and taken to England, where it advances to a baronetcy and a seat in Parliament.

Eager to find out the truth about Koeping's story of the tailed natives of Nicobar, Lord Monboddo wrote to Carl Linnaeus. He received a polite reply from the famous Swedish naturalist to the effect that he, personally, did not doubt the traveler's tales of his fellow countrymen. It was not the only time Linnaeus dealt with tailed people, a subject that obviously interested him a good deal. One of the doctoral dissertations over which he presided was "Anthropomorpha; or The Cousins of Man," which a Russian student, Christian Emmanvel Hoppius, successfully defended in 1760. One of the humanoid races presented in this thesis was *Homo caudatus lucifer*, the tailed human, which was claimed to be a "missing link" between the apes and the human races. For his description of the tailed people's habits, Linnaeus and Hoppius relied entirely on Koeping's drastic account. According to his posthumous papers, Linnaeus himself had made annotated excerpts from Koeping's fables about the tailed people, the snow-white cockroaches of the East, and the forbidden Tree of Paradise, among other subjects. He also got information about the tailed races of the East from the Dutch traveler Jacob Bondt, who had seen natives with fifteen-inch hairless tails taken to the royal court of Borneo as curiosities. The "Anthropomorpha" thesis was illustrated with a remarkable plate, showing, in a descending scale from human being to ape, the Troglodyte, the Tailed Man, the Satyr, and the Pygmy. The figure of the tailed man was taken from the zoological writings of Ulysses Aldrovandi, who had, in his turn, copied it from Conrad Gesner's *De quadripedibus*.

The famous French naturalist Georges Louis Buffon was also taken in by Koeping's tale, but the German anatomist and zoologist Johann Friedrich Blumenbach strongly condemned the Swedish seafarer's book as being "übergefüllt mit abgeschmackten Fabeln"—"packed full of repulsive tall tales." In 1799 the explorer Nicolas Fontana visited the Nicobar archipelago to investigate Koeping's old tale. He was able to verify that the "tails" observed by the Swede were only pieces of cloth, which it was the fashion among the natives to let hang down at the back.

In 1764 Linnaeus again encountered the legend, this time after a fellow Swede, the country surveyor Adolph Modéer, had written a long paper about a forest troll woman with a long tail like a horse. Colonel Anders Sparfelt of the Elfsborg Regiment had supplied him with the particulars of an extraordinary court-martial, concerning a certain Sven Joensson, whose military career had been unremarkable to say the least. After thirty years of service, he was still a private soldier. During a two hundred-mile march from Karlskrona to Gothenburg, he suddenly absconded into the woods. When his superiors caught him, the soldier confessed that for more than twenty-four years, he had regularly had carnal intercourse with a forest troll; many times, he had run away from military labor or marches in order to be with

Fig. 2. "The Cousins of Man," a plate depicting various humanoid races from the dissertation "Anthropomorpha," defended by Hoppius in 1760. From the author's collection.

her. He said that this troll had a human shape, but she was popeyed and equipped with a long, hairy tail.

The belief that such trolls frequented the deep forests of Scandinavia was still current among imaginative scholars at this time. Some of them even asserted that the trolls had been the original population of the Nordic countries, but that they had been pushed aside by the humans. Only a few individuals, such as this one, still lived in the deep forests.

Linnaeus and his fellow referees of the Academy of Science had little faith in this odd case report; they refused it for publication in the academy's transactions, and it was referred to the archives, where it is still kept 230 years later. The most likely explanation is that the woman had neither a tail nor any other troll-like characteristics and that the soldier made up the whole story in the hope of escaping punishment by claiming to have been bewitched.

It should be noted, however, that several individuals with similar hairy tails have been described in the medical literature. A so-called *naevus pilosus*, or benign hairy nevus, might be situated on the spine, and the hair is some-

Fig. 3. A speculative portrait of the Swedish woman with a horse's tail. From the author's collection.

times of considerable length. An American woman exhibited herself in various sideshows during the 1880s under the name "Lady with a Mane," and she really had a horsetail-like hairy nevus, although it was situated farther up her back. Furthermore, an eleven-year-old Indian girl with a similar "faun-tail naevus" was described in 1988. Thus, the strange story of the Swedish troll woman is not necessarily fictitious. If she really existed, it is strange that she did not simply cut off her bizarre caudal decoration with a pair of scissors, but at least it did not diminish her power of attraction for the amorous warrior.

⟋♣ While it can easily be understood why savage tribesmen of Asia and Africa and semihuman trolls from the Scandinavian backwoods might be thought to have tails, it is remarkable that a medieval myth of obscure origins ascribed such appendages to all Englishmen. According to legend, when Saint Augustine and his fellow missionaries came to England to convert its heathen population to Christianity, they were jibed and jeered by a mob of fishermen. It particularly nettled Saint Augustine that the Britons were wearing fish tails fastened to their robes. As a suitable punishment, he laid a terrible curse on this sinful island: every British child would henceforth be born with a tail. John Bale, who was bishop of Ossory at the time of Edward III, wrote in his *Actes of the English Votaries* that "for castynge of fyshe tayles at thys Augustyne, Dorsettshyre men had tayles ever after." According to another old legend, some royalists in the Kentish village of Strood wanted to insult the controversial Thomas à Becket, who was well known to be an enemy of the king. When the prelate was riding through Strood, they wantonly cut the tail off his horse. Saint Thomas avenged this low act of villainy by laying a solemn curse on the entire county: Kentish people would themselves have tails ever after. Cardinal Aeneas Sylvius Piccolomini, while on a secret mission to Scotland in 1435, mentioned the village of Strood, where the natives were supposed always to be born with tails. The epithet "Kentish Long-tails" was commonly used well into the nineteenth century. The seventeenth-century poet Andrew Marvell also mentioned this "strange tale of tails" in his *Loyal Scot:*

> *Never shall Calvin pardon's be for sales,*
> *Never for Burnet's sake, the Lauerdales;*
> *For Becket's sake, Kent always shall have tails.*

During the Middle Ages, other countries commonly used the epithet *caudatus* to deride the English. In the metrical romance *Richard Coeur de Lion*, the German emperor dismisses the king's three messengers from Cyprus with the harsh words:

> *Out, taylards, of my paleys!*
> *Now go and say your tayled king*
> *That I owe him no thing.*

In a Middle German poem, the Englishman is described in the following unflattering Latin verse: "Anglicus a tergo caudam gerit: est pecus ergo; / Cum tibi dicit 'Ave' sicut ab hoste cave!" The verse can be translated:

> *A brute beast is the Englishman*
> *For he doth bear a tail;*
> *Beware, and treat him like a foe,*
> *Even when he bids you "Hail"!*

Particularly apt to invoke the legend were the French, hereditary foes of the English, who avenged their many defeats on the battlefield by calling their adversaries *coués*. A poem dated about 1430 celebrating the victories of Joan of Arc, opens with the words, "Arrière, Englois coués, arrière!" and when the English army evacuated Paris in 1436, the inhabitants derided them by chanting "Tails! Tails!" Bishop John Bale castigated "these laisy and idle lubbers, the monkes and the priests," for exaggerating the legends of the saints to such an extent that "an Englyshman now cannot travayle in another land by way of marchandyse or any other honest occupation, but it is most contumeliously thrown in his tethe that all Englyshmen have tails." The festivities at the christening of the infant son of Mary Queen of Scots in 1566 included a play featuring actors dressed as satyrs. As they passed the English guests, they put their hands to their tails and wagged them, thereby almost causing an international incident. As late as the 1800s it happened that men from Kent were teased at rural fairs for their alleged tails by people from other counties. The English antiquary Sabine Baring-Gould remembered having it impressed on him at an early age, by a Devonshire nurse, that all Cornishmen had tails, although these might, in time, be "sat off" by people of sedentary habits.

During the heyday of the old monster medicine, the dramatic reports of tailed human races were also spiced with sporadic European cases of tailed children. One very early report appears in the chronicles of the Abbey of Hirsau, by Johannes Trithemius. In 1335 he observed a poor beggar who roamed the countryside, earning his living by dancing and showing his tail, which was the size of a man's finger, to the curious populace. In some other medieval and Renaissance cases, the child's mother was suspected of having had sexual intercourse with some tailed domestic animal. In addition to the grief of having borne a deformed child she had to endure impertinent and intimate questions from self-appointed village inquisitors. In Ambroise Paré's *Des monstres et prodiges*, published in 1573, the famous French surgeon depicted two creatures of more bizarre aspect, a dog boy and a pig man. Both were considered the hideous products of bestiality, grotesque in appearance and bearing tails. The birth of the dog boy in 1493 was considered a portent of the Pope Alexander VI's many evil deeds. According to Paré, the pig man was born in a sty belonging to Mr. Joest Dickpeert, rue Warmoesbroeck, Brussels. The original sources report that he died shortly after birth, and the illustration depicts how it was presumed he would have looked in adult years.

Other early cases of tailed infants were explained through the ancient doctrine of maternal impressions. Dean Cristopher Krahe reported a Danish case of *homo caudatus* before the Royal Society of London in 1684. Krahe believed the tail and other deformities were due to the mother's fright when she saw a soldier being wounded. Her child was therefore born with wounds on arm and leg and with an extra toe "like the *bullet* of a Pistol." After crying out two or three times, the infant died. The draftsman who illustrated the case seems to have given his imagination a free rein, depicting the child's cranial meningocele like a military cap and giving it a tail almost reaching the ground.

Isbrand van Diemerbroeck, professor of anatomy at the University of Utrecht, described another case: "The coccyx bone, it being outward bent in lenghit grows dry, becomes a Tayl, as we saw it in year 1638, in an infant born half Ell long, like the Tayl of an Ape, which was occasioned by the mother being frightened by an Ape with a tayl, after she had gone but three months. Thus Pliny tells us of some men that have woolly Tayls in some parts of India." In several nineteenth-century German cases, too, the mother was said to have been frightened by one tailed animal or another. As late as 1894 the causative event for an American *homo caudatus* was believed to be an episode in which the mother held a piglet by the tail early in her pregnancy.

In the late 1800s the interest in human tails reached its peak. The German biologist Ernst Haeckel formulated the recapitulation theory of embryology, "ontogeny recapitulates phylogeny" which held that advanced creatures, such as humans, repeated the adult stages of their ancestors during their embryonic development; in short, their embryos climbed their own

family trees. Embryology came to be seen as a useful tool to unravel the   *181*
secrets of evolution; if the child had a tail, it had reverted to a lower stage of
development. The human tail was considered to be a relic of the primate
ancestors of human beings. Many anthropologists devoted their scientific
labors to surveying the literature on this subject and classifying the tails into
subgroups; the most impressive review was written by the German Dr. Max
Bartels in the 1880 and 1884 volumes of the *Archiv für Anthropologie*, but there

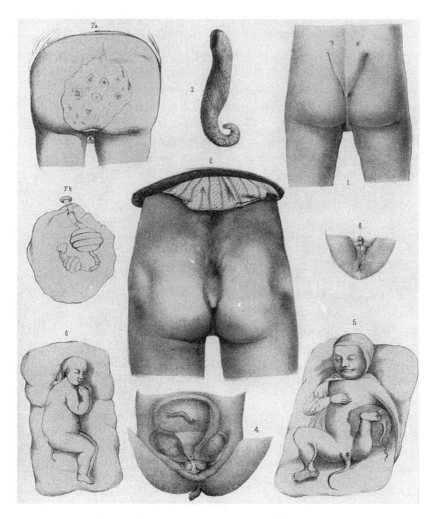

Fig. 4. A collection of tailed children from Dr. Max Bartels's 1884 review in *Archiv für Anthropologie*. From the author's collection.

182 were several other German monographs on this subject, and at least one Russian doctoral thesis.

A much-debated question was whether the human tail could contain an increased number of caudal vertebrae, as some early cases had suggested. The first such case was reported by the famous seventeenth-century Danish anatomist Thomas Bartholin, whose father had seen a *puer caudatus* on the Danish island of Fyn: this tailed boy had an increased number of vertebrae. A case report of similar age and obscurity was given by the German M. F. Lochner in the 1688 volume of the notorious journal *Miscellania curiosa*, which specialized in monstrosities, strange diseases, and macabre events. He had examined a half-starved, miserable boy of eight, whose parents were too ashamed to speak of his peculiar affliction. The boy had a considerable tail, as thick as a thumb and as long as the forefinger of a sturdy man. Dr. Lochner reported that the tail had a row of additional vertebrae inside. A third case was reported by no less a worthy than François Voltaire himself, in his *Dictionnaire philosophique*. He had seen a poor woman in Paris, who had four breasts and a long tail like a cow's. According to the surgeon Baron Percy, who had obviously also seen her, "cette femme incroyable" used to exhibit herself before the public, dressed in a fantastic costume, making a lot of money after it had become fashionable among the Paris swells to visit her and make rude jokes about her long hairy tail. A fourth case, long believed to be the best one, was reported by the Austrian Dr. Thirk, who had extirpated a caudal growth containing "a vertebra" from a twenty-two-year-old Kurd. From the description and illustration, however, it is apparent that this "tail" was in fact a so-called caudal teratoma, a neoplastic growth that might well contain bone fragments. The earlier three reports can also be justly criticized: the two seventeenth-century ones were both based only on a summary examination of the tail and appeared in less trustworthy periodicals; the extraordinary woman described by Voltaire might well have donned an artificial tail to increase her powers of attraction among the jaded Parisian *roués*.

🐦 The theory of recapitulation has been discarded by modern embryologists: the embryo develops not from primitive to advanced species but from generalized to specialized. The tail, as a general character in all vertebrates, appears rather early in the human embryo's development. At the sixth week of gestation, we all had a tail, complete with as many as ten to twelve caudal vertebrae. About half of this embryonic tail contains vertebrae, but the distal part, the caudal filament, is flaccid. During the seventh and eighth weeks of gestation, the tail disappears, and the number of vertebrae is reduced through fusion.

Many of the tailed infants described in the older and newer medical literature have had so-called true vestigial tails, which represent persistence of the outer part of the caudal filament. These true tails are only one of several dissimilar malformations that have been described as tails: there are also caudal lipomas and teratomas, as well as abnormal elongations of the coccyx. The London *Times* of 1869 had a letter from a retired officer who had once interviewed a soldier who wanted to join the Irish militia. On being asked

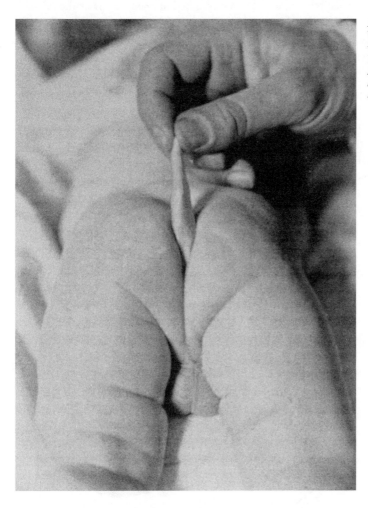

Fig. 5. A six-month-old boy with a three-inch tail, described by Dr. Harrison in the *Johns Hopkins Medical Bulletin* of 1901. From the author's collection.

184 why he had been rejected for the mounted corps, the man said, "Because I have a tail, Sir!" The military surgeon at once examined him and verified his story. The broad, unyielding elongation of the coccyx would have made his career as a cavalryman an extremely painful one!

In 1901 Ross Granville Harrison, M.D., associate professor of anatomy at Johns Hopkins University, described a remarkable case. An infant, otherwise quite healthy, had a tail that grew at an alarming rate. An inch and a half long at birth, it had grown to nearly three inches by the time the child was six months old. Interestingly, the boy would wag, or rather contract, his tail when he was irritated or when he coughed or sneezed. After the tail had been removed, Dr. Harrison examined it under the microscope. The tail was covered with normal skin and had a central core of connective tissue and fat with a normal supply of nerves and blood vessels; some rather powerful muscular strands reached through it, explaining the tail's mobility. No elements of bone or cartilage have been observed in any modern case of a human tail.

It is an open question how long a tail can get if allowed to grow indefinitely, for nowadays tails are usually amputated at an early age to spare the child and its parents anxiety and unwelcome attention. Since the tail usually grows at a fairly rapid pace, the boy described by Harrison might well have sported a ten-inch tail by the time he was college age. In the older literature, tails of six or seven inches are reported on grown subjects, and in some modern cases, quite young children have had impressive tails. The record is held

Fig. 6. Sections of the same tail, which was amputated by Dr. Harrison. From the author's collection.

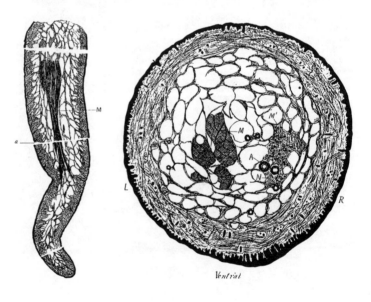

by a twelve-year-old Moi boy from Thailand, described in *Scientific American*
in 1889, who had a soft tail almost a foot in length.

The first tail amputation was performed in 1848 by the German surgeon Dr. Greve. In an Indian case, reported in 1903, the parents of a child with an impressive 3½-inch tail refused to have it amputated, since they hoped to make big money exhibiting their child. The human tail is usually, but not always, a benign stigma. A recent survey has shown that the true vestigial tail is not rarely associated with congenital midline fusion defects, so-called spinal dysraphism. In an infant I observed at the Department of Pediatrics, Lund University Hospital, in 1989 a very short true tail was associated with spina bifida occulta. Rarely, the human tail is combined with more severe malformations; an example is a child with myelomeningocele, sirenomelia, and a "pig's tail," described in Schenckius's *Monstrorum historia* of 1609.

Modern anthropologists agree that there have been no tailed human races. None of the many reports of tailed "missing links" in Asia and Africa have withstood critical reexamination. Notions of a "great chain of being" from simple to more advanced creatures made Linnaeus particularly receptive to his contemporaries' vague theories about tailed people and other odd "Cousins of Man," theories born, for the most part, of early encounters with apes. Remarkably, as late as 1861, the theologian Joseph Wolff, D.D., reported the existence of people in Abyssinia with tails long and powerful enough to knock down a horse. Similarly, in 1876 the Wesleyan missionary George Brown described a tailed tribe at Kali, off the coast of New Britain, among whom all tailless infants were immediately destroyed as an act of mercy, to spare them the ridicule and jeers of their fellows. In our own time, although a tail is no great problem from the medical point of view, and no more anomalous than a cleft palate or an extra finger, it still arouses great curiosity and interest. Some American cases have even made the headlines of the tabloid press, with as much irrelevant speculation about the child's supposed apelike characteristics as would have been likely a century ago.

Three Remarkable Specimens in the Hunterian Museum

⤳ *Hunter not only improved the practical part of surgery, but by his investigations and researches, brought it to a level with the other branches of philosophy, and connected all those surgeons who pursue it on this extended scale with the sons of science. His Collection of Preparations, which the munificence of the Government has bestowed upon this College, has enriched it beyond any other establishment for promoting the healing art.*

Sir Everard Home, from his Hunterian Oration
(1814)

JOHN HUNTER HAS RIGHTLY BEEN CALLED the Founder of Scientific Surgery, for he established methods of experimental biology and applied them to problems in practical surgery. John Hunter was as versatile as he was farsighted, and his research covered every aspect of natural history, with a special interest in zoology and comparative anatomy. His famous museum, which contained more than thirteen thousand preparations at the time of his death, is still kept in the building of the Royal College of Surgeons in London. John Hunter's interest in teratology and human anomalies is well illustrated by three specimens in the Hunterian Museum, which to this day remain almost unique.

John Hunter was born on February 13 (or 14) 1728, the tenth and last child of a Scottish countryman. He was a bad scholar and learned to read and write only with great difficulty. At the age of thirteen, he ceased to attend grammar school, at his own request; his father had died by this time, and his mother could not control the headstrong lad, who roamed the countryside at will, plundering birds' nests and catching rats and moles for dissection. Until the age of twenty, John Hunter received no further education, some months of instruction in carpentry excepted.

In 1748 John's elder brother William, who had made a brilliant career in medicine and obstetrics, employed him as an assistant in his anatomy school in London. Here, John Hunter at last found work that suited him: he assisted at dissections, helped William with his research, and procured corpses through London's sinister grave robbers, the so-called resurrection men. Some writers have implied that he actually practiced as a body snatcher himself during this period, but there is no evidence at all for this suggestion. John Hunter studied surgery under William Cheselden of Chelsea Hospital and Percival Pott of St. Bartholomew's, and later worked as a house surgeon at St. George's Hospital. William Hunter worried about his brother's ignorance in the humanities and arranged for him to spend some time at Oxford, but this scheme failed miserably. John was back in London after only a few weeks, complaining that "they wanted to make an old woman of me, or that I should stuff latin at the University." He added, "These schemes I cracked like so many vermin."

After twelve years of work at William Hunter's school and three years of service as a military surgeon during the Portuguese campaign, John Hunter settled down as a surgical practitioner in London. Initially, his success was very limited, and in order to make a living, he had to practice dentistry, a branch of science considered little above quackery by contemporary surgeons. His foremost aim as a dentist was to transplant teeth from one person to another, and he studied this odd subject with great zeal, making innumerable experiments. Hunter performed autograft tissue transplantations, embedding a cock's spur in the same bird's comb, and also heterografts, successfully (as he thought) transplanting a human tooth into a cock's comb. His experiments with the transplantation of cocks' testicles into the abdominal cavity of hens and his observation of the compensatory hypertrophy of the remainder of the testicles make him one of the earliest pioneers of endocrinology. The dental transplantations were not successful, in spite of all his efforts, and only a few of the new teeth stayed in place for any length of time. One American patient, however, was quite satisfied with his three new front teeth, which had been "donated" by some poor wretch in exchange for money and implanted by John Hunter; these teeth did good service for twelve years.

During the 1760s, Hunter built up a collection of pathological and zoological specimens which was to form the core of his museum. His collection of tooth specimens is well preserved to this day, demonstrating Hunter's skillful injection experiments, which showed how well the transplanted parts were nourished by the surrounding tissues. In 1767 John Hunter became a fellow of the Royal Society, and the year after he was appointed surgeon of St. George's Hospital. In 1771 he sold the first part of his monograph *The Natural History of Teeth* for a thousand pounds, which gave him some financial ease for the future. He was able to buy a house in London, and that year he married Anne Home, the daughter of a fellow military surgeon. She was a handsome, intellectual lady, who later kept a well-attended literary salon in London. John Hunter used to go down to shake hands with his wife's guests, but he did not stay long, preferring to spend the evening with his scientific manuscripts. In spite of their dissimilar interests, the marriage was happy, and thanks to his wife, John Hunter broadened his horizons considerably. Anne Hunter's brother Everard became one of Hunter's students, and was later promoted to be his research assistant.

As a skillful practical surgeon, John Hunter realized the limitations of the methods of his day. He likened operative therapy to an armed savage taking with force what a civilized man would obtain by stratagem. Similarly, he deplored the tendency of some contemporaries to perform unneccessary and sometimes dangerous operations for their own financial gain rather than the well-being of the patient. He used to advise his students never to perform any operation they would refuse to undergo themselves. Well aware of the danger of malignant tumors, Hunter advocated the radical extirpation of the entire growth as the only lasting cure, and several tumor specimens in his museum bear witness to the wisdom of this policy.

Another group of preparations in the Hunterian Museum illustrates his famous operation for popliteal aneurysm, which was quite common in Hunter's time. The choice for the patient was to have the leg amputated or to bleed to death when the aneurysm ruptured. John Hunter had the King's permission to experiment on the deer in Richmond Park; he particularly wanted to study the growth of their horns. Once, he ligated a buck's left carotid artery to see how the horn would be affected. The pulsations in the vessel disappeared, and the horn went quite cold. The growing antler is covered with a layer of skin well supplied with blood vessels, known as velvet, which becomes hot during the antler's period of growth. A week later, he was astonished to find that the pulsations had recurred, and the horn was warm once more. John Hunter had the buck killed, and after dissection and injection of the arteries, he could see that a system of collateral vessels had formed. He was at once aware of the importance of this finding for his patients with popliteal aneurysm. He could ligate the femoral artery in the adductor canal, far from the diseased part of the vessel. A collateral system would form, ensuring the future blood circulation in the limb. The results were excellent, and Hunter's surgical reputation was definitely made. When one of the old patients died of other causes many years later, some Hunterian surgeons dissected the corpse's leg and injected the arteries; this beautiful preparation, showing the collateral vessels with great clearness, can still be seen at the Hunterian Museum.

During the 1780s, John Hunter was one of London's leading surgeons, and his practice was very large. He had several famous patients: he treated the future Lord Byron for clubfoot with an orthopedic boot and diagnosed cirrhosis of the liver in his old friend, the famous artist Sir Joshua Reynolds, who had painted his portrait.

The great Austrian composer Joseph Haydn was a friend of Anne Hunter's. He had set several of her poems to music, and she wrote a libretto for his *Creation*, which was performed for the first time at the Hunterian Bicentennial in 1993. Haydn once consulted John Hunter for a nasal polyp; Hunter wanted to remove it, but Haydn feared the operation. Some days later, Hunter sent a message that he wanted to see Haydn on some business; when the composer arrived, two toughs seized him and forcibly seated him in an operating chair, before which Hunter's instruments were ready! The terrified Haydn managed to wrestle free and to convince Hunter that he really did not want to "undergo the happy experience of enjoying his skill," which surprised Hunter greatly.

John Hunter purchased a house with a considerable plot of land at Earls Court outside London, and went to great expense to equip it as a research station and menagerie. During the week, he would make rounds at his ward at St. George's Hospital and teach his students there, as well as taking care of his flourishing practice in Jermyn Street, but he spent the weekends in Earls Court, making physiological experiments, dissecting, and writing. Hunter liked animals and enjoyed watching their behavior. He was an expert

190   handler of both tame and wild beasts; sometimes, he amused himself by driving around in his buffalo cart. Other members of the Earls Court menagerie were pigs, porcupines, deer, snakes, bats, an ostrich, a zebra, a jackal, and some wolves. Once, Hunter's two leopards broke free and attacked some large dogs in the courtyard. The ensuing great uproar alerted Hunter himself; he dashed into the melee, grasped the leopards by the scruffs of their necks and dragged them back to their cages.

    John Hunter's contacts with sailors and research expeditions gave him many opportunities to obtain specimens of rare and exotic animals. In all, he dissected more than five hundred different species, and this variety gave him unique insights in comparative anatomy. Along with the Dutchman Pieter Camper, Hunter was the first to demonstrate that birds' air cells communi-

Fig. 1. John Hunter's portrait, by his friend and patient Sir Joshua Reynolds (1785). Reproduced by kind permission of the Royal College of Surgeons of England.

cate with the bones; he realized that the capacity of respiration was increased by these auxiliary respiratory organs. The Hunterian Museum illustrates his ambition to classify animals and to observe how structure was related to function in their organs. A large series of preparations concerns the adaption of the structure of the gastrointestinal canal to the animals' conditions of life. One of the most remarkable experiments concerned a seagull, which Hunter fed with corn for a year. The muscular layer of its stomach could be seen to be considerably thickened to enable it to digest this unaccustomed nourishment. He later tried the same experiment on an eagle, which took this kind of food only with much reluctance; one day, to Hunter's chagrin, the dissatisfied bird broke its chain and flew away.

Another section in the Hunterian Museum deals with the skeletal changes involved in syphilis. In Hunter's time, syphilis and gonorrhea were considered to be different manifestations of the same sexually transmitted disease. Furthermore, it was believed that both were curable by large doses of mercury. In his *Treatise on the Venereal Disease*, John Hunter maintained that the two diseases were one and the same, although his clinical observations tended to separate them. One section in this book has been taken to imply that Hunter, in order to study the progression of symptoms in detail, had inoculated his private parts with pus from a patient, thereby contracting both gonorrhea and syphilis. This assertion is to this day uncritically repeated in almost every paper about Hunter, but his latest biographer, Mr. George Qvist, has established that it has little foundation in the original sources and that it is unlikely that Hunter was a "martyr of science."

By 1773 Hunter experienced his first attack of ischemic heart disease, and during the 1780s he was constantly troubled by severe angina pectoris. Nonetheless he did not in any way slacken his pace. In 1786 he published *Observations on Certain Parts of the Animal Oeconomy*, his major work on zoology and comparative anatomy, and the following year he received the Copley Medal from the Royal Society. In 1790 he became surgeon general of the army. Hunter was well aware that his attacks of angina pectoris could be brought on by anxiety or irritation, and he used to say that his life was in the hands of any scoundrel who tried to vex him. During a board meeting at St. George's Hospital on October 16, 1793, one of those present flatly contradicted him. According to Everard Home, it was the surgeon Robert Keate. Hunter rose from the table to deliver a furious riposte but collapsed with a groan and staggered into the adjoining room, where he died shortly after. Everard Home and Hunter's other pupils performed the autopsy, which showed severe generalized arteriosclerosis.

Hunter's executors, his brother-in-law Everard Home and his nephew Matthew Baillie, met with many difficulties in their efforts to preserve his museum intact according to his testamentary wishes. Hunter had invested almost seventy thousand guineas in building up his great collection and had few other assets. During his final years, he had employed more than fifty servants, animal keepers, and préparateurs. All but his amanuensis, William Clift; a préparateur, Robert Haynes; and his housekeeper, Mrs. Adams, were sacked after his death. The Hunters' carts, horses, books, furniture, and paintings were sold to pay the debts. Through the good offices of the king, Anne Hunter was given a modest widow's pension. When Everard Home approached the prime minister, hoping to persuade the government to purchase the entire Hunterian Collection, William Pitt is reported to have an-

Fig. 2. John Hunter with a giant and two dwarfs. An original drawing from the extra-illustrated copy of Jessé Foot's *Life of Hunter* in the Wellcome Institute Library, London. Reproduced with kind permission of the library.

swered, "What! Give £20,000 for bottles — we want the money to buy gunpowder!" After some lean years, when the entire museum was kept in storage with young William Clift as the sole custodian, the government finally bought the museum in 1799 and entrusted it to the Corporation of Surgeons.

An integral part of the collection was Hunter's voluminous posthumous papers. During his entire life, he had been most reluctant to publish the results of any investigation that had not been concluded to his complete satisfaction, and large bundles of manuscripts and books of drawings and descriptions of animals, containing observations of immense value, were turned over to his executors. Furthermore, Hunter had not been able, during his lifetime, to publish the results of his extensive researches in geology and paleontology, since they disagreed completely with the Bible's teachings about Creation and the Deluge. Whereas the biblical chronology taught that Earth had existed for only about six thousand years, Hunter considered it "many thousand centuries" old. Furthermore, whereas the traditionalists claimed that all species of animals and plants had been independently created and had remained unchanged ever after, Hunter's studies of fossils had made him aware that animals could become extinct and that the species now living were related to those that had died out.

The Irish Giant

In John Hunter's time — and indeed to this very day — the most spectacular exhibit in his museum was the skeleton of the Irish Giant, Charles Byrne. This celebrated personage was born in 1761 in the village of Littlebridge, Ireland. Although his mother was rather unflatteringly described as "a stout woman with a strong voice," both of his parents were of normal stature, nor was he himself a particularly large infant. The village gossips attributed his great increase in height — already in youth, he "grew like a cornstalk" — to his conception atop a very high haystack. When he was a boy, his schoolfellows teased him unmercifully. He suffered from joint pains; his stamina was generally low; and he was constantly dribbling and spitting. When he was in his teens, he was discovered by the showman Joe Vance, who exhibited him at local fairs and markets with considerable success. Later, the Irish Giant was quite a success in Edinburgh, where he astonished the night watchmen by lighting his pipe at one of the streetlamps. In the Old Town, he had great difficulty in getting up and down the narrow stairs, having to crawl on hands and feet. In 1782 Joe Vance took the Irish Giant to

Fig. 3. A watercolor by T. H. Shepherd of part of the Hunterian Museum, circa 1860. Reproduced by kind permission of the Royal College of Surgeons of England.

London, where he was installed in a disused cane shop next door to the late Cox's Museum at Spring Gardens. The newspapers spared no superlatives in describing "the modern Living Colossus," and although it cost as much as half a crown to see him, the Giant did not lack interested spectators. Newspaper advertisements claimed that he was eight feet four inches tall, thus far exceeding both the famous German Maximilian Miller and the Swedish Giant Daniel Cajanus. "In short," according to a contemporary pamphlet, "the sight of him is more than the mind can conceive, the tongue express, or pencil delineate, and stands without parallel in this or any other country. 'Take him for all in all, we shall scarce look on his like again' — Shakespeare." Charles Byrne's mounted skeleton is only seven feet, seven inches, however, which makes it more probable that he was about seven feet, ten inches in height, taking into account the intervertebral substance. Giants have always exaggerated their heights!

Initially, Charles Byrne and his clever manager did quite well in London. Several contemporary diarists and letter writers mention visits to the famous Irish Giant; one of the most perceptive was Sylas Neville: "Tall men walk considerably under his arm, but he stoops, is not well shaped, his flesh loose, and his appearance far from wholesome. His voice sounds like thunder, and

Fig. 4. The Irish Giant, a drawing by Thomas Rowlandson (1782). Reproduced by kind permission of the Royal College of Surgeons of England.

he is an ill-bred beast, though very young — only in his 22nd year." In early 1783 the fickle public began to crave new attractions. Furthermore, at the rumor of Byrne's financial success, several other giants tried their luck in the metropolis. Byrne was especially annoyed that a gigantic fellow country-man, Patrick Cotter, also called himself the Irish Giant. Other competitors were the Gigantic Twin Brothers named Knipe, who were also natives of Ireland; they even claimed distant kinship with the Byrne family. Joe Vance lowered the entrance fee to one shilling and moved to a less expensive apart-ment in Cockspur Street, but to little avail. Charles Byrne wasn't particularly clever, and his health and judgment were further impaired by his habit, ac-quired at an early age, of drinking large quantities of gin and whiskey every day. Shows often had to be canceled because of the Giant's drunkenness and lack of punctuality. For a while, he acted as porter at St. James's Palace, but he soon tired of this post. One night in April 1783 Byrne went on an ex-tended drinking spree, visiting a pub called the Black Horse, among others. The next morning, the sottish Giant was aghast to find that his entire savings of £770 in banknotes, which he had been unwise enough to carry with him in his pocket, had been stolen. This hard blow lowered his spirits consider-ably, and he sought comfort in new bouts of drinking. His habitual drunk-enness made him weak and sickly, and in May 1783 he appeared to be a dying man.

Byrne was not yet dead when many London surgeons who had seen and admired him when he was on exhibition began to wrangle over who would get the corpse for dissection. A newspaper article stated that "the whole tribe of surgeons put in a claim for the poor departed Irish Giant, and surrounded his house just as Greenland harpooners would an enormous whale." John Hunter was the most eager of all. He had employed a man named Howison to follow Byrne around in case the invalid giant died suddenly.

Charles Byrne was aware of Hunter's schemes. Like most eighteenth-century people, he had a great fear of being dissected. Since Hunter and his colleagues were allied to the gangs of body snatchers who prowled the churchyards nightly in search of freshly buried corpses, no grave would be safe for him. The Irish Giant promised some fishermen his last remaining savings if they saw to it that he was buried at sea, where the medical men could not reach his corpse. This confidential deal "leaked" to the press, how-ever, and the newspapers averred that the body hunters were "determined to pursue their valuable prey even in the profoundest depth of the aquatic regions; and have therefore provided a pair of diving bells, with which they hope to weigh hulk gigantic from its watery grave." When Charles Byrne finally died, the fishermen took his body to the Downs, where it was to be sunk in twenty fathoms of water, but Hunter and Howison managed to catch up with them. They got the fishermen drunk, bribed them with £500, and the Giant's body was instead delivered to Hunter's house at Earls Court, where he boiled it in a great iron kettle and prepared the skeleton for his museum. One of the Giant's boots and one of his slippers and gloves are also there, as well as several drawings of him. Another, perhaps more truthful account states that Hunter and Howison simply bribed a London undertaker to obtain the body.

The Irish Giant's skeleton was one of John Hunter's most prized possessions, and he pondered much on what might be the etiology of Byrne's abnormal growth. This question was not resolved until 1909, when the famous Harvey Cushing, who was the first to propose that acromegaly and gigantism resulted from hypersecretion of growth hormone from the hy-

Fig. 5. An etching by John Kay of Charles Byrne and the two gigantic Knipe brothers, from Kay's *Edinburgh Portraits* (1885). From the author's collection.

pophysis, wrote to the conservator of the Museum of the Royal College of Surgeons of England, Sir Arthur Keith, inquiring whether the Giant's cranium might be sawed open for examination. Keith found great enlargement of the pituitary fossa due to the pressure of a pituitary tumor that had grown upward and forward. Later radiographic examination of the skeleton confirmed that the cause of the Irish Giant's growth was a pituitary adenoma producing growth hormone. As is typical of this condition, Byrne was weak and sickly in spite of his great stature; some have speculated that Goliath also was a sufferer and that the weakness associated with the disease accounts for his ignominious defeat at the hands of David. Today, such a pituitary adenoma can be dissected out, using a transsphenoidal approach and modern microneurosurgical techniques. Had the Irish Giant lived today, surgery could have been performed on him at a relatively early age. He might never have been a national celebrity or gained immortality in a museum; instead, he might have led a longer and happier life.

The Two-Headed Boy of Bengal

Another remarkable specimen in the Hunterian Museum is a child's skull of most bizarre aspect, that of the famous Two-Headed Boy of Bengal. In 1790 John Hunter's principal assistant, Everard Home, rightly described this boy as "a species of *lusus naturae* so unaccountable, that, I believe, no similar instance is to be found upon record." Home never went to India to see the boy, but he collected several descriptions and drawings of him from laymen. The Two-Headed Boy was born in May 1783 in the village of Mundul Gait in Bengal; his parents were poor farming people. Immediately after the child was delivered, the midwife, greatly terrified by its strange appearance, tried to destroy the infant by throwing it into the fire; the boy was saved from the flames, though he had burns on one eye and one ear on the upper head. The parents soon realized that they could make money by exhibiting their child in Calcutta, where he attracted much attention. Such large crowds gathered to see him that his parents had to cover him up between shows. In this way the unhappy boy spent his short and miserable life. His emaciated and sickly appearance, Home thought, was caused by his being covered by sheets most of the time. The Two-Headed Boy's fame soon spread all over India, and several noblemen and civil servants had him exhibited in their houses. One of these, Colonel Pierce, described him in a letter to the president of the Royal Society, Sir Joseph Banks, who later gave the letter to Everard Home.

The two heads were of the same size, and covered with black hair at their junction; the upper head ended in a necklike stump, which one observer likened to a small peach. When the boy cried or smiled, the features of the upper head were not always affected, and their movements seem to have been purely reflex: a pinch in the cheek produced a grimace, and when it was given the breast, its lips attempted to suck. The lower head and body were quite normally developed, but a number of anomalies were noted in the parasitic head: the corneal reflexes were absent and the eyes' reaction to light

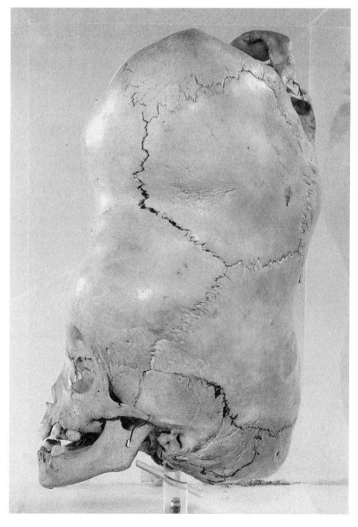

Fig. 6. The double skull of the Two-Headed Boy of Bengal, as it can be seen at the Hunterian Museum (specimen no. P 1535). Reproduced by kind permission of the Royal College of Surgeons of England.

200   was weak. No pulsations could be felt in the temporal arteries of the parasitic head, but its superficial facial veins were seen to be well filled with blood. The ears were malformed, and the tongue small. The lower jaw was rather small, but capable of motion. The secretion of tears and saliva was plentiful. When the child slept, the eyes of the parasitic head could be observed to be open and moving. When the boy was first awakened, all four eyes moved in the same direction, but normally, the two heads' eye movements were independent.

When the boy was four years old and in general good health despite his emaciation, his mother left him one day to fetch water. When she returned, her son was dead from the bite of a cobra. Several scientific amateurs offered to purchase the corpse, but the religious parents refused such desecration; they buried their child near the Boopnorain River, at the town of Tamluk. The grave was later plundered by Mr. Dent, the East India Company agent for salt in Tamluk. He dissected the putrefied body and kept the skull, which he gave to Captain Buchanan of the same company. The captain brought the skull to England, where he gave it to his friend Everard Home. When Mr. Dent dissected the heads, he noted that the brains were separate and distinct, each enveloped in its proper coverings. The dura mater of each brain adhered firmly and contained many large vessels, supplying the nutrition to the upper head. In his examination of the double skull, Home noted that the halves were nearly of the same size. No septum of bone existed between the two brains. The natural skull was quite normal, but the parasitic one

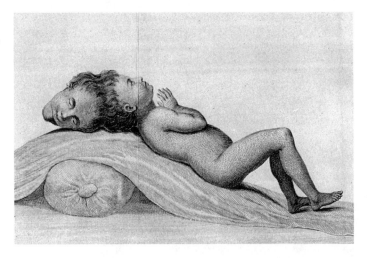

Fig. 7. Mr. Smith's drawing of the Two-Headed Boy, from Sir Everard Home's *Lectures on Comparative Anatomy,* vol. 4 (1823). From the author's collection.

was imperfect in a number of particulars, and its facial bones were generally
smaller.

Everard Home much regretted that "men of observation" never got
the opportunity to examine the boy. The two brains and their influence on
the "intellectual principle" fascinated him. Neither Home nor John Hunter
speculated about how the Two-Headed Boy should be classified in the system of teratology. They do not seem to have noted the similarity between the
boy and the symmetrical conjoined twins of the craniopagus type, which are
connected at the crowns of the heads. The junction may be more or less
extensive. In partial craniopagi, the crania are intact or simply fused and the
brains are distinct and complete. In total craniopagi, the two brains are en-
cased in a common cranium, and the brains are sometimes malformed. Craniopagi had been described several times in the old annals of teratology. The
famous twins of Worms, frontal craniopagi born in 1495 and depicted in
several crude woodcuts of the time, lived for ten years. During the eighteenth century, several well-attested cases were reported. Today, at least
eighty bona fide cases of craniopagus twins have been described; this type

Fig. 8. Two very well made watercolor drawings of the Two-Headed Boy of Bengal, by a certain Mr. Dent. These were used by Everard Home in his 1799 paper on the boy, and engraved
for the *Philosophical Transactions* of the Royal Society. Reproduced by kind permission of the
Royal Society, London.

represents 6 percent of all double monstrosities, making the incidence 0.6 per million births. The surgical techniques for separating craniopagus twins have improved during the last decades, but the operation is still a most uncertain and difficult venture, especially in cases with total conjunction and malformations of the brains and their vascular supply.

In his *Traité de tératologie*, published in 1836, the famous French anatomist Isidore Geoffroy Saint-Hilaire was the first to note the similarity between the Two-Headed Boy of Bengal and the craniopagi. He quoted Home's case report at length, and criticized the British surgeon for his lack of teratological knowledge. Saint-Hilaire suggested that the Two-Headed Boy belonged to a special group of parasitic craniopagi, which he called *épicome*, and he cited a similar French case, described by the Liège surgeon M. Vottem in 1828. Here, the parasite had been much larger, possessing incompletely developed arms and a rudimentary spinal column. The monstrous newborn was seen to gasp for breath and move slightly for about half an hour; after its death, the mother was told that the baby was normal but stillborn, in order to spare her anxiety. In all, nine cases of this teratological type, which is called *craniopagus parasiticus* in the English system and *épicome* in the French one, have been reported to this day. Home's case was the first of these, and the only one capable of prolonged extrauterine life. The most likely teratogenetic mechanism is that they begin as a symmetrical craniopagus, but one twin loses its contact with the umbilical vesicle, and those parts not directly supplied with blood through anastomoses from the healthy twin do not develop. This theory is supported by the finding in several cases that the necklike stump on top of the parasite's head contains a small sternum, clavicles, and some underdeveloped ribs, as well as a rudimentary heart, pharynx, and lungs. The spinal column has been normal down to the lower cervical portion, and the lower vertebrae have been fused and highly hypotrophic.

Although the Two-Headed Boy was the only craniopagus parasiticus to live much past birth, there have been several instances of viable parasitic conjoined twins when the parasite was located elsewhere. Of these, Ambroise Paré's famous case of a man with a parasitic head emerging from the epigastrium has been doubted by some teratologists, but James Poro, born in Genoa in 1686, was afflicted in almost the same way. He was exhibited in London in 1714 before Sir Hans Sloane, who commissioned a portrait of him. The celebrated Lazarus-Joannes Baptista Colloredo, who was born in 1617, also deserves mention. He carried quite a large parasitic twin, consisting of head, trunk, arms, and one leg, from the epigastrium; he himself was quite strong and healthy and the father of several children. Colloredo was well described by Thomas Bartholin, who saw him in Copenhagen in 1645 at the age of twenty-eight. As in Home's case, the parasite could be awake while Lazarus slept. It moved its hands, ears, and lips when touched, but its respiration was weak and it did not take any nourishment. Bartholin described Lazarus as comely in person and morals and with the polite bearing of a courtier.

Had the Two-Headed Boy lived today, it would have been relatively easy to resect the parasite and enable the boy to lead a near-normal life, at least if the legal status of the parasitic head, which seems to have given some signs of independent life, could be resolved in this unprecedented case. The operation would have been facilitated by the separateness of the brains. The skull defect was relatively small, and the tissues of the parasite could be used to cover it. Thus, had he been born two hundred years later, the Two-Headed Boy of Bengal might not have had to live as a miserable object of exhibition.

The Sicilian Fairy

After John Hunter's death, his brother-in-law, Everard Home, advanced swiftly in the surgical world. He published several influential works on practical surgery and urology and established a lucrative private practice in London. In 1808 Home became sergeant-surgeon to King George III. Two years later, he was called in to attend the king's fifth son, the duke of Cumberland, who had been attacked while he slept by his servant, Sellis. The cowardly footman, who later cut his own throat, had given his master several deep sword cuts about the head and neck. After the duke had recovered from this assault, Home was rewarded with a baronetcy by his royal friend and patron, the Prince of Wales. As a scientist, Home was as versatile as Hunter himself, but he lacked his great teacher's creativity and genius. Sir Everard Home published more papers in the Royal Society's *Philosophical Transactions* than anyone else in the history of that famous body, but at least in his later years, these many contributions were of varying quality. Some of his contemporaries even hinted that Home used undue influence on his friend and patient Sir Joseph Banks, the president of the Royal Society, to get his manuscripts printed. In 1821 Sir Everard Home became the first president of the Royal College of Surgeons, and in the same year, he was appointed surgeon of Chelsea Hospital. Like John Hunter himself, Home was something of a connoisseur of "monsters" and medical curiosities. He had written on the Two-Headed Boy of Bengal and had published a lengthy paper on hermaphro-

204   dites in the *Philosophical Transactions*. When the body of a sea monster washed ashore on the island of Stronsay in 1808, some believed it to be the great sea serpent, but Home correctly identified it as a large basking shark. In 1822 he made an anatomical examination of an alleged dried mermaid being exhibited in London and announced that it consisted of the top half of a baboon and the bottom half of a large fish.

        In early April 1824, at the height of his distinguished career, Sir Everard Home went to see Caroline Crachami, the Sicilian Fairy, a dwarf who had

Fig. 9. The bust of Sir Everard Home by Sir Francis Chantry. Reproduced by kind permission of the Royal College of Surgeons of England.

recently been taken to London for exhibition by a certain Dr. Gilligan, an Irishman who claimed to be the child's father. The entrance fee was a shilling, but to be allowed to lift her up and examine her more closely, the spectators had to pay an extra shilling. The animated little girl apparently made a great impression on Home. At the age of nine, she was only 19½ inches in height and 11¼ inches around the waist. She walked without any support, but without much confidence. She was attracted to glittering objects and liked dressing up in fine clothes; she had a taste for music and was happy to see some visitor who had previously showed her kindness. The little girl knew enough English to express herself rather fluently and gave relevant answers to questions asked her by the audience.

Another, less appealing view of the exhibition is given in the *Memoirs of Charles Mathews* by his wife, published in 1839. When the visitors entered, the Sicilian Fairy sat on a low throne "in seeming mockery of regal state." Her voice was very thin and high-pitched, and the showman had to repeat all her utterances for the benefit of the audience. Mrs. Mathews noted that he added "many particulars not mentionable to ears polite" in order to amuse them. The tall Dr. Gilligan had dressed up in strange garb in an attempt to look like an Italian, but he could not disguise his thick Irish brogue. When he named his place of birth, Charles Mathews, a time-worn habitué of London low life who delighted in various monster shows, asked him "whether it was Palermo in the *County of Cork* where he was born?" The Irishman gave

Fig. 10. Caroline Crachami is exhibited, a watercolor sketch by John Augustus Atkinson. Reproduced by kind permission of the Yale Center for British Art, Paul Mellon Collection.

him an arch leer and admitted, "Och! I see your honour's a deep 'un! Sure, you're right, but don't *peach*!" To bribe Mathews to secrecy, Gilligan offered to let him take the child up and examine her for nothing, thus saving a shilling, but Mathews declined the rogue's offer with scorn and later freely denounced the man's imposture to his friends.

Sir Everard Home went to see Caroline Crachami several times, and he soon became friendly with his unscrupulous colleague Gilligan, who realized that he could make use of the influential baronet's interest in his tiny protégée. They commissioned a tailor to make her a richly embroidered dress, and Sir Everard took her to Carlton Palace and introduced her to his royal friend, King George IV, who the newspapers remarked, "expressed high pleasure at her appearance."

The Fairy's presentation at court heralded a time of greatly improved economic and social circumstances for Dr. Gilligan and his alleged daughter. The odd ceremony was reported in many of the newspapers, and it became fashionable to visit the exhibition. Soon, the Fairy had as many as two hundred paying visitors daily, many of them so fond of her that they came several times. The *Morning Chronicle* reported that since her presentation at court, "she had eclipsed most of the distinguished fashionables; the morning calls of the Royal Family, the Nobility, the Foreign Ambassadors, and the highest members of the Faculty, and others of rank and fortune, have frequently exceeded two hundred, and the number of presents she has received exceeds all precedents." Dr. Gilligan soon took better lodgings, on Duke Street, and also rented a suite of rooms on Bond Street as a new exhibition hall. The newspapers saw through his attempt to pose as an Italian and hinted that he had also lied about the birthplace of his alleged daughter, suggesting that "the phenomenon was really a native of Ireland."

Dr. Gilligan soon had an exhibition pamphlet printed, titled *Memoirs of Miss Crachami, the Celebrated Sicilian Dwarf*, in which he stated that she was born in Palermo on November 15, 1815, the daughter of Signor Louis Emmanuel Crachami. Like his wife, this gentleman was by profession a theater musician. He was healthy and of normal stature; his wife, it was slyly remarked, "is justly considered a fine woman." They had four other children, all of normal stature. At birth, Caroline Crachami had weighed only one pound and measured between seven and eight inches long. Her parents had showed her privately to the duchess of Parma and other noble curiosity seekers, but they did not consent to having her exhibited for money before the populace. In the pamphlet, no reason was given for their sudden change of mind in this respect, nor was it explained why the Fairy had been taken to England.

One of Caroline Crachami's most fervent admirers was the newspaper man William Jerdan, who wrote extensive accounts of this "wonder of wonders" in his column. Never in his life had he seen anything like her: "Only imagine a creature about half as large as a new-born infant; perfect in all parts and lineaments, uttering words in a strange, unearthly voice, understanding what you say and replying to your questions; imagine, I say, this figure of about nineteen inches in height and five pounds in weight, — and you will have some idea of this most extraordinary phenomenon." Caroline's large nose and somewhat microcephalous cranium made her look older than

Fig. 11. The portrait of Caroline Crachami by Alfred Edward Chalon. Reproduced by kind permission of the Royal College of Surgeons of England.

nine years, and her almost adult bodily proportions added to this impression. Her hands and feet were slender and graceful, and she moved her arm "with all the motions and grace that are found in the same member of a lovely woman." Caroline did not like people to measure her or examine her too minutely, and whenever doctors were mentioned in the conversation, she "doubles her filbert of a fist, and manifests her decided displeasure." In order to get her measurements, Jerdan invited her to his house, gave her a ring and a large doll, and treated her to a meal of biscuits and diluted wine, which she relished very much. Thus he won her over, and she let him measure her; her height was 19½ inches, the length of her foot was 3⅛ inches, and the length of her forefinger 1⅞ inches. Her head had a circumference of 12⅜ inches, and her waist measured 11¼ inches. Caroline Crachami conversed animatedly and observantly and expressed many opinions, likes and dislikes, impatience, enjoyment, mirth — the last prevailing.

Some weeks later, the newspapers announced Caroline Crachami's sudden death. For some days, she had coughed and appeared unwell, but her illness did not dispose her showman to make the exhibition schedule less fatiguing. After receiving more than two hundred visitors on June 3, 1824, toward the evening languor overtook her, and she died in the carriage on the way home to the Duke Street lodgings. Sir Everard Home suspected that the cause of death was consumption, since she had suffered from a "hacking cough." The Fairy's death bitterly grieved her many admirers in London, and in the newspapers, Gilligan was criticized for callously exploiting her. The poet Thomas Hood, who had probably seen her alive, gave her a fitting epitaph in his *Ode to the Great Unknown*:

> *Think of Crachami's miserable span!*
> *No tinier frame the spark could dwell in*
> *Than there it fell in —*
> *But when she felt herself a show, she tried*
> *To shrink from the world's eye, poor dwarf! and died!*

A week after the Sicilian Fairy's demise, there was an extraordinary scene in the Magistrate's Court at Marlborough Street. Her real father, Louis Emmanuel Crachami, employed as a musician at the Theatre Royal in Dublin, approached the magistrate, F. A. Roe, for advice on how to reclaim his daughter's dead body. It turned out that he had consulted Dr. Gilligan in Dublin about his daughter's ill health. The medical man, who was much struck by her extraordinary appearance, had considered the Dublin climate much too cold for her delicate health and recommended a stay in London.

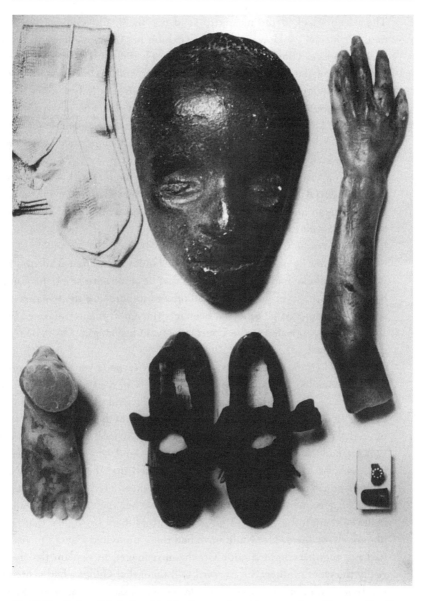

Fig. 12. The Sicilian Fairy's ring, thimble, shoes, and stockings, together with casts of her face, arm, and left foot and ankle (specimen no. 227, Human Osteology). Reproduced by kind permission of the Royal College of Surgeons of England.

The parents lacked the means to send her there, but Gilligan offered to take her himself, if he was allowed to exhibit her for money in order to pay their expenses. Thus having secured the hapless child, Gilligan took her on an exhibition tour of Liverpool, Birmingham, and Oxford to earn enough money for lodgings in London. The parents were kept in ignorance of Gilligan's nefarious proceedings, and it was only through the newspapers that they learned of their child's death. The Fairy's grief-stricken mother wanted to see her child one last time before burial, and Crachami traveled to London to bring her corpse back.

He first visited the Bond Street exhibition room, which was just being closed down. At the Duke Street lodgings, he met the landlord, furious because Gilligan and his entourage had absconded without paying the bill of twenty-five guineas. The doctor had vanished without a trace, leaving behind the costume Caroline Crachami had worn at court, as well as a "state bed" given her by a wealthy admirer. The landlord said that while the Fairy was alive Gilligan had already bragged that he could earn as much money on her after death; several surgeons had offered to purchase the remains, and a sum of five hundred pounds had been mentioned. Crachami, alarmed by the sinister news, appealed to the Court to prevent his daughter from being sold for dissection.

Roe lacked the authority to order the body turned over to Crachami, but he recommended that an inquest into the cause of death be held by a coroner's jury. In the meantime, Crachami and a friend of his went around to several London anatomists, trying to find out whether the body had been offered for sale. At Joshua Brookes's anatomy school, they found out that Gilligan had tried to sell the Fairy's body for a hundred guineas, but Brookes had declined to buy it. The landlord suggested that they visit Sir Everard Home's house in Sackville Street, since he remembered that this gentleman had been a frequent visitor at the exhibition. When they arrived, Home, who believed them to be Gilligan's hired hands, told them brusquely that he had no money for them. He would have had these uninvited visitors turned out, had not the distraught Signor Crachami managed to explain that he was really the child's father. Sir Everard told them that Gilligan had visited him on June 7 to offer the corpse for sale. Home passed this offer along to the Royal College of Surgeons, and he obviously expected that they would purchase it for the museum, since it is recorded in the Museum's Donation Book that Home himself took Caroline Crachami's body there in a box the same day. Since Gilligan wished to leave London as soon as possible, it was agreed that he would send a messenger back to collect the money voted to him by the college.

Louis Emmanuel Crachami was appalled at this final evidence of Gilligan's villainy. He asked Home if he could have one last look at his child. Sir Everard gave the distraught Italian a cheque for ten pounds to calm him down and also an entrance pass to the museum, but the unfortunate father had another shock waiting for him there. Home's principal assistant, William Clift, and his pupils had been so eager to examine the body that they had already started, and the dissection apparently was well advanced. With tears flowing, Crachami embraced the dismembered corpse, and it was only with difficulty that his friends could persuade him to leave it. The newspapers had much to say about these touching scenes, and the Sicilian Fairy's tragic fate became headline news. The last word of the scandal from the press was that Signor Crachami left London for Dublin to convey "the dreadful intelligence" to his wife and that Gilligan, who had netted fifteen hundred guineas on the exhibition, was rumored to have fled to France.

Although the surgeons had promised Crachami that the dissection was to proceed no further, Home ordered the Sicilian Fairy's skeleton prepared and mounted. He also managed to preserve her pearl ring, thimble, and tiny stockings and shoes. Home was not noted for generosity, and it is unlikely that Dr. Gilligan ever received any pecuniary reward for his villainous body snatching. The ten pounds he had paid Crachami, Home apparently considered the purchase price of the body. He donated the skeleton and Fairy memorabilia to the museum in his own name, and it was set up next to that of the Irish Giant, whose enormous boot would easily have contained the lot.

In a paper published in the Royal Society's *Philosophical Transactions,* Home used Caroline Crachami's case to illustrate the terrible effects of a "maternal impression." When, in 1815, Signora Crachami was almost four months gone with child, she was sleeping in a caravan in the baggage of the duke of Wellington's army in France. In a violent storm, a small monkey, which traveled on the roof of the caravan, bounded in through the window and crept under her skirts to keep warm. Later, she awoke from the pressure of the monkey and wanted to scratch herself but came upon the monkey's head. The startled animal bit her fingers and seized hold of her loins, and the poor woman went into fits. A miscarriage was expected, but she carried her child to term. The result seemed a foregone conclusion to supporters of the ancient belief in maternal impressions.

Sir Everard Home's own description of the Sicilian Fairy's autopsy is regrettably brief, but fortunately, William Clift's handwritten autopsy report is still extant. Except for a small amount in the posterior parts of the ocular orbits, there was no fat in any part of the body. The liver was healthy but the bile ducts and gall bladder almost as large as those of an adult. The lungs were almost entirely covered with irregular whitish spots and very much tuberculated throughout. Furthermore, the inguinal and mesenteric glands were numerous and much enlarged. The urinary bladder was extremely distended with urine, but both kidneys seemed healthy. No further examination of these parts was undertaken, and Home concluded that the monkey's grip on the loins of the mother had caused this affliction of the bladder when the child was in utero.

The Sicilian Fairy was for many years one of the most spectacular exhibits in the Hunterian Museum. While the tiny skeleton, standing between those of the giants Charles Freeman and Charles Byrne, was a familiar sight to generations of English surgeons, it was considerably less well known on the Continent. The first after Home to express any opinion about the cause of her condition was the teratologist and surgeon Sir Hastings Gilford. In a long paper on dwarfism published in 1902, he suggested that she was a victim of "ateleiosis," a newly defined condition characterized by proportional arrest of growth and development. During the twentieth century, many more cases of proportional dwarfism were published. In 1960 Professor H. P. G. Seckel of the University of Chicago hypothesized an autosomal recessive syndrome of intrauterine growth retardation, postnatal dwarfism, and microcephaly with a bird-headed profile, a large, beaklike nose, and a receding chin. He considered Caroline Crachami one of the earliest and most extreme cases of "Bird-headed dwarfism." Her birth weight is the lowest ever recorded for this syndrome. An even earlier historical case was Nicolas Ferry, better known as Bébé, who was the court dwarf of King Stanislaus Leczinski of Poland from the age of four until his death in 1764 at the age of twenty-one. His skeleton, which is preserved at the Musée de l'Homme in Paris, resembles Caroline Crachami's in several respects. Both have very light, thin bones, hypoplastic lower jaws, and craniums characteristic in shape and with prominent nasal bones.

Almost all the cases of the Seckel syndrome have been mentally retarded, often severely so, although there is no definite evidence that Caroline Crachami was particularly feebleminded. Both Gilford and Seckel claimed that certain points in Home's description, such as her quickness of sight, attraction to bright objects, and pleasure in music and fine clothes, would point toward mental retardation, but since these observations concern a nine-year-old girl, their judgment is a trifle severe. The newspaper man William Jerdan, who appears to have studied her at least as closely as Home, considered her much like any normal-size child of the same age. All accounts of her agree that she was quick-witted and amusing; although Italian was her native tongue, she learned to speak English with alacrity. William Clift's autopsy report confirms that she suffered from severe tuberculosis, as was correctly diagnosed during life by Sir Everard Home. This disease might account for some of her bodily infirmities and might have been responsible for variations in her mental state due to exhaustion.

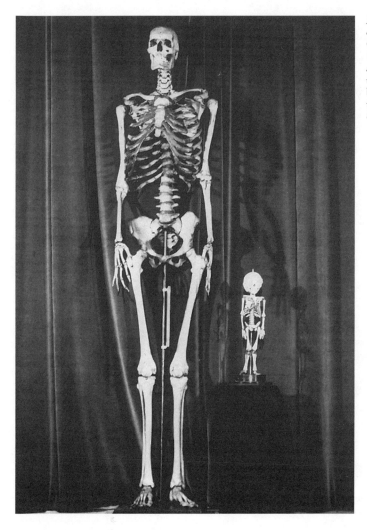

Fig. 13. The skeletons of Charles Byrne and Caroline Crachami (specimen nos. 223 and 227, Human Osteology). Reproduced by kind permission of the Royal College of Surgeons of England.

214   During the last twenty years, Seckel's classification of osteodysplastic primordial dwarfism has been challenged by several leading geneticists. Today, it is clear that several subgroups exist within this entity. Some lately published cases have had facial similarities to Caroline Crachami, and also similar dysharmonic retardation of bone age, but large differences in intelligence, bodily incapacity, and birth weight. Indeed, there has been no case exactly like hers in the annals of this syndrome. Thus, the Sicilian Fairy is memorable not only for her strange and tragic life story, but also as an early and nosologically interesting case of a rare genetic disorder.

The Hunterian Museum Today

The description of the Sicilian Fairy was one of Sir Everard Home's last scientific triumphs. He died in 1832, and his later years were embittered by an unsavory scandal, in which his long-suffering assistant, William Clift, reported to the trustees of the Hunterian Collection that Home had burned John Hunter's invaluable posthumous papers after having plagiarized them in his own publications for many years. Although these accusations were

Fig. 14. The Hunterian Museum today. Reproduced by kind permission of the Royal College of Surgeons of England.

probably exaggerated, Home lost the confidence of his colleagues, and to this day he has been considered the villain of the Hunterian tradition. The part he played in Caroline Crachami's tragic fate is certainly nothing to be proud of.

John Hunter was so far ahead of his time that the full extent of his genius was unknown to most of his contemporaries. He was buried in the vaults of Saint Martin-in-the-Fields after what was little more than a pauper's funeral. In 1859 the surgeon and natural historian Frank Buckland decided to find his coffin. After looking at nearly three thousand coffins among the noxious effluvia in the vaults, he found it in excellent condition, identifiable by the inscription on a brass plate on the lid. John Hunter was reinterred at Westminster Abbey among the nation's most revered people. He was finally acknowledged as the greatest of his profession and a symbol of the rise of British surgeons from simple barbers and sawbones to respected men of science. "Hunter made us gentlemen," said Sir James Paget in his Hunterian Oration. "When he entered surgery it was a trade; when he died it had become a science." John Hunter is today one of the most revered figures in the history of medicine.

The diligent and dedicated William Clift was the first conservator of the museum, and in 1842 he was succeeded by the famous Sir Richard Owen. During this time, the museum was the leading establishment of its kind in England and received many foreign visitors. During the 1880s and 1890s, many valuable additions to the collection were made by another dedicated conservator, Professor Charles Stewart. After him Sir Arthur Keith remained conservator until 1935. In May 1941 the building of the Royal College of Surgeons was hit by several German incendiary bombs, followed by a direct bomb hit in one of the museum rooms. Not less than 60 percent of the Hunterian Museum was destroyed, including many of Hunter's invaluable original preparations. The specimens saved from the debris were put into storage, and even during the war, the Royal College of Surgeons was making plans to reopen the museum. John Hunter's original plans were followed as closely as possible, and new specimens were provided only when an integral part of one of the series in comparative anatomy or physiology was found to be missing. In 1963 Princess Margaret and Prime Minister Harold Macmillan were in attendance when the museum was reopened. The collection still reflects Hunter's unique achievements, and is more didactically arranged than in his own time. The skeletons of the Irish Giant and the Sicilian Fairy stand next to each other in the same glass case, and the double skull of the Two-Headed Boy of Bengal on one of the shelves.

The Strange Story of Julia Pastrana

'Twas a big black ape from over the sea,
    And she sat on a branch of a walnut tree,
And grinn'd and sputter'd and gazed at me
    As I stood on the grass below;
She sputter'd and grinn'd in a fearsome way,
And put out her tongue, which was long and grey,
And hiss'd and curl'd and seem'd to say
    "Why do you stare at me so?"

Her ears were pointed, her snout was long;
Her yellow fangs were sharp and strong;
Her eyes — but surely I must be wrong,
    For I certainly thought I saw
A singular look in those fierce brown eyes;
The look of a creature in disguise;
A look that gave me a strange surmise
    And a thrill of shuddering awe.

216

T HESE ARE THE FIRST TWO STANZAS OF "Pastrana," a remarkable poem by the English civil servant Arthur Munby, which was published in *Relicta*, his final collection of poems, in 1909. After the long and detailed description of the grotesque, apelike creature observed by the poem's narrator in the garden of a small German hotel, he describes his dinner in the hotel restaurant, where he is fascinated by a well-dressed lady of peculiar aspect:

> *Sure, I remember those bright brown eyes?*
> *And the self-same look that in them lies*
> *I have seen already, with strange surprise,*
> *   This very afternoon;*
> *Not in the face of a woman like this,*
> *Who has human features, and lips to kiss.*
> *But in one who can only splutter and hiss —*
> *   In the eyes of a grim baboon!*

In the ludicrous conclusion of the poem, the hotel waiters catch the ape-woman after a desperate struggle and drag her howling from the room in a large net, to the relief of the frightened narrator. When an admirer wrote to Arthur Munby to find out how the inoffensive civil servant could have thought of such a bizarre subject, Munby replied that the poem had a certain background in reality. In 1857 he had seen a peculiar monster, called "Julia Pastrana, the Baboon Lady," being exhibited for money, and the sight had made a lasting impression on him, bearing fruit in this thirty-two-stanza poem, published fifty-two years later.

When she was exhibited to the public in the United States, Canada, and Europe in the 1850s, and long after her death in 1860, Julia Pastrana was one of the most famous human curiosities of her time. She was a Mexican Indian woman with excessive hairiness over large parts of the body, as well as an overdevelopment of the jaws which gave her an apelike visage. After her death in childbirth, her body and that of her son were embalmed by a Russian professor. Julia Pastrana's husband, who had been her impresario, continued to exhibit the mummies of his hairy wife and child. After his death, they were taken over by other showmen, finally ending up at a Norwegian fairground. The name Julia Pastrana was a byword in large parts of Europe, even long after her death. In 1964 the Italian movie director Marco Ferreri produced the film *La donna scimmia* (The woman-ape) which was directly based on her strange life story.

Julia Pastrana's Origins

The early history of Julia Pastrana is veiled in mystery. She appears to have been born in 1834 and to have belonged to a tribe of so-called Root-Digger Indians, who inhabited the western slopes of the Sierra Madre in Mexico, facing the Gulf of California. According to the exaggerated accounts in the contemporary exhibition pamphlets, an Indian woman named Espinosa had become separated from her tribe in 1830, and was believed to have drowned. Six years later, however, some cowboys found her in a cave.

Fig. 1. A Russian drawing of Julia Pastrana, probably made in 1860. Reproduced by kind permission of the St. Petersburg Circus Museum, Russia.

She told them that she had been captured by a party of hostile Indians, who had imprisoned her in the cave, but no human beings could be found nearby. The place where she was found was said to be "a region of country abounding in monkeys, baboons, and bears." The woman was carrying a two-year-old girl, and she "professed to love this child dearly, though she disclaimed being its parent." With her husband, Espinosa took care of the child and had her christened Julia Pastrana. Some years later, Julia's supposed mother died, and she was sent to the nearest town. The pamphlets do not mention whether her body was already covered by hair, but this is likely to have been the case, for she was soon after taken into the family of Pedro Sanchez, governor of the state of Sinaloa at the foot of the Sierra Madre, who wanted to study her as a curiosity. She was brought up to be a serving girl and stayed in the governor's house until April 1854, when she decided to return to her Indian tribe after being ill used. While she was on her way home, an American named M. Rates spotted her and persuaded her to accompany him to the United States, to be exhibited for money.

In December 1854 Rates and Julia Pastrana traveled to New York via New Orleans. The "Marvellous Hybrid or Bear Woman," as she was first called, was exhibited at the Gothic Hall, 316 Broadway. A contemporary newspaper account describes her: "The eyes of this lusus natura [*sic*] beam with intelligence, while its jaws, jagged fangs and ears are terrifically hideous. . . . Nearly its whole frame is coated with long glossy hair. Its voice is harmonious, for this semi-human being is perfectly docile, and speaks the Spanish language." In New York, Julia Pastrana attracted a good deal of attention both from the public and from doctors and scientists. She was first examined by Alexander B. Mott, M.D., who declared her to be "one of the most extraordinary beings of the present day," a hybrid between human and orangutan. With her new showman, J. W. Beach, Julia went on to several other American cities. In Cleveland she was seen by a professor named S. Brainerd, M.D., who examined her hair under the microscope and found that she had "no trace of Negro blood." It was his view that her hairy skin and protruding jaws "entitle her, I think, to the rank of a DISTINCT SPECIES." Mott's and Brainerd's certificates were added to the ostentatious exhibition pamphlet, which stated that all the Root-Digger Indians of Pastrana's tribe were as hairy as she and that their features bore "a close resemblance to those of a bear or Orang Outang." In Baltimore Julia was invited to a military ball, where each of those present was introduced to her and shook her hand. Some of the more daring military gentlemen even danced the waltz

220   and schottische with her. When the exhibition went on to Boston, she was shown at the Horticultural Hall during a fair. In the three-page exhibition pamphlet, there is a crude drawing of her, titled "Hybrid Indian! The Misnomered Bear Woman." Pastrana was also exhibited at the Boston Natural History Society, and Mr. Samuel Kneeland Jr., former curator of comparative anatomy for this society, declared in an affidavit that she certainly was entirely human and "a perfect woman, performing all the functions of her sex."

Fig. 2. A Russian drawing of Julia Pastrana from a poster published in late 1859. Reproduced by kind permission of the St. Petersburg Circus Museum, Russia.

"The Nondescript" in London

In July 1857 an advertisement in the London newspapers announced that "a Grand and Novel Attraction" had come to the metropolis: "Miss JULIA PASTRANA, the NONDESCRIPT, from the United States and Canada, where she has held her levees in all the principal cities, and created the greatest possible excitement, being pronounced by the most eminent Naturalists and Physicians the Wonder of the World." At this time, "nondescript" was a term freely applied to strange animals and monsters from beyond the seas. In an expansive, twelve-page exhibition pamphlet, a good deal of information was given about her early history. Julia's new impresario, Mr. Theodore Lent, also put advertisements into many of the daily newspapers, in which no superlatives were spared in describing her various attractions. Not only was her body covered with hair, but her skull was also covered with a fleshy substance, on which grew a cumbrous mass of straight, glossy, jet-black hair. Lent feared that the well-informed English public would not believe the previously reported story that all the Root-Digger Indians were as hairy and apelike as Julia. Therefore, he described her as "a hybrid, wherein the nature of woman predominates over the ourang-outang's." Julia's mother had strayed into a mountain region devoid of human beings — Digger Indians or other — but full of apes, baboons, and bears. This ignores, of course, the fact that apes and baboons are not to be found in America.

The pamphlet writer considered Julia Pastrana far superior to the wretched Digger Indians, who were described in the most derogatory terms. These Indians were no taller than four feet and weighed only from 80 to 90 pounds, but Julia stood four feet six inches tall and weighed 120 pounds. Furthermore, she lacked their spiteful and obstinate disposition and was good-natured and sociable. She could speak both English and Spanish, as well as her native tongue, and had learned to sew, cook, wash, and iron during her sojourn in the household of Governor Sanchez. She loved to travel, her health was excellent, and she was as eager to learn new things and retained knowledge as avidly as an eight-year-old child. She was kindly and affable during the shows and willing to submit to any examination to demonstrate that her extraordinary appearance was not an imposture. She was, said the pamphlet, always cheerful and perfectly contented with her situation in life. In short, Mr. Lent, who must have written or at least supplied material for the pamphlet, seems to have considered her a model freak, a house-trained monster who behaved well in front of the audience. The pamphlet stated without irony that she did not see the necessity of making money, but

222 "there are hopes that she will acquire in time the money-getting faculty, equal to that of the rest of the family of man." Lent himself did not lack this faculty: "The Nondescript" attracted a good deal of publicity in the newspapers, and the exhibition soon became one of the most popular in town. During the shows, Julia entertained the audience by singing romances in English and Spanish, and dancing the highland fling, the schottische, and other "Fancy Dances." Most of the accounts of Julia agree that she was a good dancer and sang well in a mezzo-soprano voice. After the entertainment, those who wished could get to know her better: "Miss Julia is pleased when the Ladies and Gentlemen ask her Questions, and examine her pretty Whiskers, of which she is very proud."

In his memoirs the English showman van Hare recalls visiting Julia Pastrana in private, accompanied by his famous American colleague P. T. Barnum. Her face was covered with a thick veil, which she did not remove until Mr. Lent, who had invited them, came into the room. The exhibition of

Fig. 3. A nineteenth-century copy of an 1857 photograph of Julia Pastrana, the original of which is kept in the Buckland collection at the Royal College of Surgeons of England. From the author's collection.

"the Nondescript" naturally attracted much attention from men of science as well. In the *Lancet* Dr. J. Z. Laurence described her as four feet six inches in height, thick-set and well-proportioned in body, and "intelligent and quick." Her body was hairy except for the palms of her hands and the soles of her feet, "especially on those parts that are ordinarily covered with hairs in the male sex." The hair was very thick and jet-black, and had no disposition to curl, not even the long beard and whiskers. He added that her breasts were remarkably well-developed and that "she menstruates regularly."

The naturalist Frank Buckland, who loved to frequent the various monster shows in which London abounded, also saw and spoke to Julia Pastrana when she was exhibited and later described her in one of his many books of essays on natural history. "Her features," he said, "were simply hideous on account of the profusion of hair growing on her forehead, and her black beard; but her figure was exceedingly good and graceful, and her tiny foot and well-turned ankle, *bien chaussé*, perfection itself."

In *A Terrible Temptation*, the novelist Charles Reade had one of the characters describe "Julia Pastrana, a young lady who dined with me last week, and sang me 'Ah perdona,' rather feebly, in the evening. Bust and figure like any other lady, hands exquisite, arms neatly turned, but with long silky hair from the elbow to the wrist. Face, ugh! forehead made of black leather, eyes all pupil, nose an excrescence, chin pure monkey; briefly, a type extinct ten thousand years before Adam." Reade had probably seen Julia Pastrana when she was exhibited in London.

Although it does not seem that Charles Darwin himself saw Julia when she was exhibited, he certainly took an interest in her and heard about her from his friend Alfred Russel Wallace. Darwin later described her in his book *The Variation of Animals and Plants under Domestication*, with the words: "Julia Pastrana, a Spanish dancer, was a remarkably fine woman, but she had a thick masculine beard and a hairy forehead."

Travels on the Continent

After the lucrative tour of London and the provinces, Mr. Lent and Julia Pastrana left England late in 1857 and went to Berlin, where she attracted much publicity. The German authorities discouraged degrading monster shows such as this one, but Lent managed to secure permission to exhibit her by emphasizing that Julia Pastrana only performed as a singer and dancer. When they reached Leipzig, she played the leading role in a play called

224 *Der curierte Meyer*, written especially for her and performed at the Kroll Theater. In the play a stupid German dairyman falls in love with a woman who always wears a veil; when her suitor was not on stage, Julia Pastrana would lift the veil to the great amusement of the audience. This burlesque fun continued for several acts, before Julia showed her face to the dairyman, who was instantly "cured" of his infatuation. The German police had spies pres-

Fig. 4. A fine circus poster depicting Julia Pastrana dancing, probably made in Germany in 1857. Reproduced by kind permission of the St. Petersburg Circus Museum, Russia.

ent at the first night, however, and the theater was closed after only two performances of the play, on the grounds that it was immoral and obscene. Furthermore, some German obstetricians strongly objected to the public exhibition of Julia Pastrana, for fear that pregnant women might miscarry at the sight of her, or even have children exactly like her through a "maternal impression." Thus, Lent had to make do with having Julia dance Spanish pepita-dances and sing popular songs from England and Mexico. Soon after the scandal of the closed play, the weekly magazine *Gartenlaube* published an extensive interview with Julia Pastrana, illustrated with a fine original drawing by the artist Herbert König. The portrait shows her hairy face and ears and her impressive mustache and beard. Her nose was very broad and flat, and her ears uncommonly large. Her lips were exceedingly thick, and according to the accompanying description, the tongue was large and shape-

Fig. 5. A drawing of Julia Pastrana by the German artist Herbert König (1857). From a poster in the author's collection.

less. The newspaper man was impressed by her fluent conversation. She spoke of her stage triumphs during her tour of America and England. She said that she had had more than twenty offers of marriage during her American tour, but she had accepted none because these suitors were not rich enough. The interviewer suspected that her showman had told her to repeat this story in order to attract wealthy admirers.

The profits from Julia's German tour were excellent, and soon several rival showmen and circus managers wanted to put her under contract. Lent managed to forestall their designs on his protégée by marrying her, thus securing her services in a more permanent way. Naturally, he had no scruples about continuing to exhibit her, and early in 1858 they went on to Vienna. The exhibition was as popular as ever, and several medical men also attended. Herr Sigmund, professor of anthropology in Vienna, published a brief report on the transatlantic phenomenon. He had never seen anything remotely like her and considered her type of hairiness unique. Julia Pastrana was very unwilling to submit to the professor's *Manualuntersuchung* (physical examination) of her body, but Lent, to whom she was touchingly devoted, persuaded her. Sigmund also had the opportunity to speak to her in private, and he found that she was certainly no semihuman monster trained to perform a few tricks, as the pamphlets maintained. In fact, she was intelligent, happy, and pleased with her position in life. In view of her illiteracy, Sigmund was impressed by the range of her knowledge on various subjects.

Julia Pastrana never left her apartment during the day because her manager thought that her drawing power would be diminished if she were seen by nonpaying spectators, but in the evenings she often put on a heavy veil and went to the circus with Lent. The German circus man Hermann Otto, who also appears to have met her during her stay in Vienna, agreed that Julia was clever and eager to learn and that she was good-hearted and a good judge of people. But he said that her abnormal appearance gave her much pain, and she felt ashamed to be shown as a freak of nature. When her impresario instructed her to carry a flower in her hand during the shows and to wear an elaborate headdress, it was only to emphasize the difference between her and the rest of her sex. According to Otto, Julia Pastrana was not illiterate but an avid reader who knew the world only through her books. A popular Austrian singer and actress named Friederike Gossmann knew Julia Pastrana well and visited her many times, being deeply touched by her tragic fate.

After her sojourn in Vienna, Julia Pastrana and Lent toured Germany together with Rentz & Hinné's troupe of equestrian performers, and there is

MISS JULIA PASTRANA
L indescriptible (de Mexico)
La plus grande curiosité de monde.
Elle parle anglais et espagnol, elle chante et danse.

MISS JULIA PASTRANA
the nondescript, (from Mexico)
The greatest natural curiosity in the world
she speaks english and spanish, sings and dances.

MISS JULIA PASTRANA
Die Unbeschreibliche(von Mexico)
Die grösste Natur-Seltenheit der Welt
Sie spricht englisch und spanisch, singt und tanzt.

MISS JULIA PASTRANA
Die Unbeschreibliche(von Mexico)
Die grösste Natur-Seltenheit der Welt
Sie spricht englisch und spanisch, singt und tanzt.

Fig. 6. Four drawings of Julia Pastrana in various costumes, with captions in three languages. From an 1858 circus poster, reproduced by kind permission of the Harvard Theatre Collection, Houghton Library.

228   some evidence that she actually performed some acrobatic tricks on horse-back. She also sang, danced, and played the guitar and harmonica. In late 1858 Julia visited Poland, and seems to have caused quite a sensation in Warsaw. The dancing shows were apparently regarded as a slightly obscene and scandalous public amusement, as evidenced by the caricature drawing of her by the artist Kostrzewski.

Fig. 7. A caricature drawing of Julia Pastrana, made in 1859 by the Warsaw artist Kostrzewski. Reproduced by permission of the British Library, London.

Julia Pastrana's Death

In late 1859, Lent and Julia Pastrana went on to Moscow. As always, the show was quite a success, and they made more money than ever. After some months, Julia noticed that she was pregnant. The doctors feared a difficult childbirth because of the narrowness of her pelvis, and she was attended by several trained obstetricians from the Moscow Accoucheur-Institut, under the direction of one M. Trettenbacher. Labor began on March 20, 1860. Because of the size of the infant, obstetrical forceps had to be resorted to, and some lacerations could not be avoided, but Julia Pastrana was delivered of a living boy at four o'clock that afternoon. She had hoped that the baby would be like his father, but his body was hairy all over and deformed in the same way as hers. The newborn boy soon fell into a state of asphyxia, but he was revived by the medical attendants; the child finally died after having lived only thirty-five hours. The distraught mother herself died on the fifth day after the delivery, according to the romantics from a broken heart but according to the pathologist from "metro-peritonitis puerperalis." As usual, Lent was highly aware of the commercial possibilities, and it seems

Fig. 8. A drawing of Julia Pastrana being presented to a very fat man; the individual to the far left is probably Mr. Lent. Reproduced by kind permission of the St. Petersburg Circus Museum, Russia.

that a crowd of titled spectators even visited Julia's deathbed and heard her purported last words: "I die happy; I know I have been loved for myself." The macabre Mr. Lent was of course highly put out, having lost his prime source of income, but he managed to retrieve some further capital by selling the corpses of his wife and child to Professor Sukolov of Moscow University. The professor took the bodies to his Anatomical Institute and embalmed them, using a new method of his own devising.

Before embalming them, Sukolov made a thorough examination of the corpses. Julia Pastrana's body was four feet six inches tall, and she weighed 112 pounds; the head was rather large in proportion to the body and united to it by a short, thick neck. The neck, arms, and legs were covered with hair, but the hairy growth terminated on the back in an angle, the apex of which pointed downward. The corpse's head was "unprecedented in the history of the development of the human body." All parts of it except the eyes were covered with thick, black, bristly hairs. The beard was quite thick, with the whiskers hanging down on both sides like two plaits. The broad, flat nose had wide apertures, and the lips were very thick and tightly drawn. The gums were excessively thick, with a number of excrescences. The body of the infant was also examined: it was nineteen and a quarter inches long and weighed about eight pounds. The skin was of a whitish color, and the head formed much like the mother's — the nose flattened and the lips very thick. The infant's forehead was covered with thick hair down to the eyebrows, and the neck, shoulders, and back were also totally covered with hair.

Professor Sukolov embalmed the bodies by injecting a decay-arresting mixture of secret composition; they were photographed during this process. The child's body was quite fresh, and there was no difficulty in embalming it. The mother's body was a more difficult matter, since it had decomposed a good deal. It had to be injected several times over a long period before it was finally free of the odor of decay. The whole process took six months. The parts of the mummy that had started to decompose had turned gray, but the rest of the skin retained its dusky yellow hue. The breasts were somewhat wrinkled, but the face was little changed in the process, except that the eyes had sunk in somewhat, the lips were thinner than in life, and the hypertrophied gums had also shrunk. According to Frank Buckland, there was "great rascality connected with the whole business" of the embalming, but he was "not at liberty to mention the particulars." It is certain that Lent and the professor drew up a contract, and it is possible that the corpses were shown to visitors or otherwise indelicately treated.

The Embalmed Female Nondescript

Professor Sukolov placed the embalmed bodies of Julia Pastrana and her child in the anatomical museum of the University of Moscow, where they became quite an attraction. The scientists of a zoological-paleontological expedition that was sent out from the University of St. Petersburg could not resist stopping at the Moscow anatomical institute to see the famous mummies and describe them in the annals of the expedition. Meanwhile, when the macabre Mr. Lent saw how extremely lifelike the mummies were, he decided to get them back into his own custody. It is recorded that Lent had to present a certificate of his marriage to Julia, attested by the American consul, to establish his claim to the return of the mummified bodies. It seems likely, therefore, that legal means had to be used to persuade Sukolov to give them up. According to another version, Lent had to pay the professor the equivalent of £800 for the mummies, having previously sold the corpses to him for £500. Lent hoped to get permission to exhibit the mummies in Rus-

Fig. 9. Julia Pastrana's mummy, photographed during the early 1860s. From a print in the author's collection.

232 sia, but to his great disappointment, the authorities refused. The resourceful impresario soon had an alternative scheme ready, and in February 1862 the world's most famous bearded lady again appeared before her many admirers in London. The price of entrance was a shilling, considerably lower than during Julia's lifetime, but Lent could keep the exhibition open for longer hours. Furthermore, the mummy of the child had also been taken to London, and was set up next to its mother on a small pedestal. Julia Pastrana's mummy was dressed in an elaborate Russian dancer's dress of her own making, and the child in a peculiarly wide sailor suit. The remarkable exhibition of "the Embalmed Female Nondescript" was much noted in the popular press, and the *Lancet* commented: "All interested in the methods of preserving the dead will do well to examine the result of this curious and completely successful system." Frank Buckland also went to see her, and at the sight of the curious object on its small platform, exclaimed, "Julia Pastrana!" The showman, probably Lent himself, assured him that it really was the famous Nondescript, returned to her London admirers, and he allowed Buckland to examine the mummies closely. The senior taxidermist of the British Museum, Mr. Bartlett, who accompanied Buckland, had never seen mummies as well made as these. Frank Buckland gives an excellent description of Julia Pastrana's mummy:

> The figure was dressed in the ordinary exhibition costume used in life, and placed erect on the table. The limbs were by no means shrunken or contracted, the arms, chest, &c. retaining their former roundness and well-formed appearance. The face was marvellous; exactly like an exceedingly good portrait in wax, but it was *not* formed in wax. The closest examination convinced me that it was the true skin, prepared in some wonderful way; the huge deformed lips and the squat nose remaining exactly as in life; and the beard and luxuriant growth of soft black hair on and about the face were in no respect changed from their former appearance.

When the novelty of the Embalmed Female Nondescript had begun to fade, the mummies were lent to an English traveling museum of curiosities. In 1864 they were taken on a tour of Sweden, together with a German anatomical museum; a wax mold of Julia Pastrana was also on exhibition. The abominable Lent was now searching for some new "artist" to exploit, and when he passed through Karlsbad he learned that a young woman there had a considerable beard. Although she was kept locked in the garden of her family's town house, Lent managed to become acquainted with her, and some weeks later he asked her parents for her hand in marriage. After some time for consideration, her grim father permitted the marriage, but only if Lent promised never to exhibit his wife for money. The cunning showman agreed to these demands, but soon after the wedding, he took away his wife's shaving tools and made plans for a grand tour through Europe. To exploit her predecessor's notoriety, he announced his new bearded wife as Miss Zenora Pastrana and stated in the exhibition handbills that she was Julia's sister. Sometimes, Lent even tried to start rumors that they were one and the same person. Again, he succeeded beyond all reasonable expectations. For more than ten years, they signed lucrative contracts with Europe's finest circuses and gave private performances for several royal families. Initially, the mummies were also on tour with them, but not for long; perhaps Zenora did not want to see her mummified predecessor's ghastly, stiff grin and feel the cold stare from the black glass eyes. At any rate, the mummies were lent to the Präuscher Volksmuseum at the Prater in Vienna; the mother of the proprietor Hermann Präuscher agreed to pay Lent a yearly rent of 320 taler. This Volksmuseum was a large fairground where various shows of giants, dwarfs, and deformed people were an important part of the entertainment.

The Restless Mummies

In the early 1880s Mr. Lent and Zenora retired from show business and went to St. Petersburg, where they purchased a small waxworks museum. After the lucrative tours with his two bearded wives, Lent was a wealthy man, and he allowed the Präuscher Museum to keep the mummies for a considerable annual payment. In 1884, however, he was struck by "acute weakening of the brain," and began to dance in the streets and to tear up the bank notes and stock certificates he had earned in such a peculiar way and throw them into the river Neva. He was taken to a Russian insane asylum, where it is unlikely he survived long. In 1888 Zenora Pastrana left Russia for Munich, picking up the two mummies, which, like the rest of Lent's estate, were now her property, on the way. In November of that year, she exhibited herself and Julia Pastrana's mummy before the Anthropological Society of Munich, possibly to scotch the rumors that she and the mummified celebrity were in fact the same person, as Lent had tried to make people believe. A professor of anatomy named Rüdiger commented on the very skillful mummification of Julia Pastrana's body. About the character of Mr. Lent, the Munich anthropologists commented that the bizarre American

Fig. 10. Mr. Lent's second bearded wife, "Zenora Pastrana." She suffered from secondary hirsutism, quite a different condition from that of poor Julia. From a print in the author's collection.

certainly must have had "eine ganz eigenartige Vorliebe für solch haarige Schönen" — "A quite remarkable fondness for such hairy beauties."

In 1889 Zenora gave the mummies to a German impresario named J. B. Gassner, who ran an "anthropological exhibition" in Munich. The hirsute Frau Lent herself retired permanently from show business, settled in Dresden, and married a man twenty years her junior. Herr Gassner exhibited the mummies at various German fairs, and in 1895 he took them to a large circus convention in Vienna, where he sold them. They changed hands several times during the next twenty-five years, and in 1921, they were bought by the Norwegian fairground owner Håkon Jaeger Lund, who was building up a "chamber of horrors" at his large amusement park near Oslo. The mummies of Julia Pastrana and her son were part of this chamber of horrors for many years, along with many other bizarre preparations, such as half a man's corpse in a glass box, a human skin including the scalp, and many monsters in glass bottles, as well as a large collection of wax molds illustrating various diseases. As far as the scientific world was concerned, the mummies had disappeared. In 1926 the German circus historian Alfred Lehmann made a radio talk on Julia and Zenora Pastrana, and he ended it by asking his listeners if anyone knew the present whereabouts of the mummies. Although many people replied with stories of having seen the two bearded celebrities at some stage of their careers, no one knew where Julia Pastrana's mummy was stored.

Early in 1943, during the German occupation of Norway, the medical director of the German forces in Oslo, Dr. Müller, ordered the wax molds and other exhibits in the Lund chamber of horrors confiscated, the wax to be used in the war production. But the fairground owner's son, Hans Jaeger Lund, managed to save the preparations by suggesting that the wax molds be taken on an extended tour of Sweden, the profits to benefit the treasury of the Third Reich. Dr. Müller agreed to this scheme, and the old fairground director went to Sweden with three railway cars full of preparations, including the mummy of the Apewoman, as Julia Pastrana was now called. For some years, he toured Sweden exhibiting these objects at fairs and markets, but the German treasury does not seem to have benefited much from this odd scheme.

In 1953 Håkon Jaeger Lund stored the chamber of horrors, including the mummies of Julia Pastrana and her child, in a warehouse in Linköping, Sweden. It was soon rumored that the warehouse was haunted by apelike, monstrous creatures, and some daring youths broke into it to investigate. They were so horrified by the sight of Julia Pastrana's mummy staring at

Fig. 11. A German advertisement, probably from the 1890s, of the mummy of "the most interesting woman in the world," as Julia Pastrana was called. From the private collection of Mr. Bjørn Lund, Oslo.

them from its dusty, cobwebbed sedan chair that they fled the building in panic. The most intrepid of them later returned to take a photograph of the mummy, the first for more than eighty years. It was still dressed in its dancer's outfit and lashed to its Russian sedan chair, but the ravages of time are clearly visible. The jewelery was gone, the elaborate coiffure destroyed, and the extraordinary beard and whiskers existed no more. The son's mummy had also lost a good deal of its hair, as well as its elaborate costume.

In 1954 Håkon Jaeger Lund died and his son Hans took over the fairground. In 1959 the chamber of horrors was exhibited at a large agricultural fair in Oslo, where it was seen by more than seven hundred thousand people. The weather was very hot, and the sight of the monsters, the hideous wax figures, and the "stuffed apewoman" proved too much for many a Norwegian countryman. The ambulances had to pay several daily calls to pick up visitors who had fainted. In the early 1960s the market for the chamber of horrors was less lucrative, and it was permanently stored in a warehouse outside Oslo.

Julia Pastrana's Last Tour

In September 1969 the name Julia Pastrana hit the headlines again after more than a hundred years. Judge Hofheinz, a wealthy American collector of curiosities, had found out that the mummies still existed and that they probably were kept in Sweden or Norway. He was very eager to purchase them for his collection and hired the Swedish circus director Trolle Rhodin to try and find them. There were many responses to an advertisement in the daily newspapers offering a reward for information. Lots of people reported seeing the mummy when it was exhibited in the 1950s. After a week Hans Jaeger Lund also contacted Rhodin, but he considered the American's offer of ten thousand dollars too low. To force up the price, he threatened to take the mummies on tour again, after cleaning them up with the vacuum. He also claimed that several other American collectors and showmen were interested in purchasing the mummies and even wanted people to believe that the Mexican government had offered a very high price because it was planning to erect a mausoleum for Julia Pastrana in Mexico City. These bizarre proceedings attracted much attention from the Swedish and Norwegian press, and there were several articles about the valuable mummies. Lund showed one journalist an old Russian document (now lost), dated 1860 and including an affidavit that the corpses of Julia Pastrana and her son had been made into genuine mummies. Lent had probably obtained this document from

238   Professor Sukolov; some early writers mention that such a document was on show next to the mummies at the Präuscher Museum. Although the American collector increased his offer several times, Lund still declined to sell the mummies; instead, he exhibited them in both Sweden and Norway in 1970. They were quite a public attraction thanks to the publicity in the newspapers and the objections against such a degrading monster show from various religious organizations. Then, the American collector suffered a stroke, and

Fig. 12. The mummies of Julia and her son on exhibit in Malmö, Sweden, in January 1970. Photographed by Mr. Rolf Olsson. Reproduced by permission of Scandia Photopress, Malmö.

Lund lost his buyer, as well as the five hundred thousand dollars Hofheinz was allegedly offering just before he died.

In 1971 the mummies were exhibited all over Norway, and they were also rented to a Danish fairground for a short time. The next year they were rented by an American traveling amusement park called Million Dollar Midways, which took them on tour throughout the United States. They were placed in a cage made of unbreakable glass to prevent theft, and the whole contraption was carried in an enormous circus van. Although several American states closed the exhibition after a few days, judging it to be considered immoral and degrading, the enterprise was a monetary success. In 1973 Lund again planned to have Julia Pastrana exhibited in Norway, but there was a public outcry against it, and the church objected to the use of a dead body in such a sordid commercial way. The acting bishop of Oslo, Reidar Kobro, demanded that the mummies be confiscated and buried by the Norwegian church. The witty Hans Jaeger Lund replied that if the Norwegian clergy wanted to bury mummies, they could start in Egypt! When his plans were thwarted in Norway, he simply rented the mummies to Björkman's Tivoli of Arboga, and they were taken on a grand tour of Sweden in the

Fig. 13. The circus trailer with Julia Pastrana's mummy, photographed in Köping, Sweden, in 1973. From a photograph in the author's collection.

summer of 1973. During this tour, Julia Pastrana attracted as much attention as the most popular entertainers and pop stars; in small towns, people
thronged to see the world's only mummified apewoman, still in the Russian
sedan chair, which was exhibited in the large van with its glass cage. The
mummy of the little son was standing beside that of the mother like a parrot,
with the feet in their black boots nailed to a pedestal. As when Julia Pastrana
was on show in London in 1857, the exhibition handbills declared her to be
a hybrid between human and ape.

When the Björkman amusement park arrived at Hudiksvall, a small
town in central Sweden, the local board of health closed the exhibition of
the Apewoman, and a petition was sent to the Swedish Home Office asking
that exhibition of dead bodies be prohibited by law. The exhibition of Julia
Pastrana's mummy was banned in Sweden, and the Norwegian Home Office
also banned any further public show of the mummies, citing a law from 1875,
under penalty of confiscation. Julia Pastrana and her son had made their last
tour, and their caravan was stored at the fairground's winter quarters near
Oslo. Three years later, Hans Jaeger Lund died and his son Bjørn Lund took
over the fairground. He did not want to exhibit the mummies in public, but
he felt that they should be kept for posterity. In August 1976 the fairground
was broken into, and the burglars forced the lock of the mummies' caravan.
Julia Pastrana's dress was torn open and the child mummy's arms and lower
jaw were knocked off; it was also taken off its pedestal and thrown into a
ditch outside. The damaged child's mummy was later eaten by mice, and
Julia Pastrana now stood alone in her glass cage in the large van. During the
summer of 1979, the fairground was again broken into, and Julia Pastrana's
mummy was among the objects missing. It was presumed that vandals had
stolen and destroyed it.

The Mummy Is Rediscovered

In February 1990 a Norwegian detective magazine revealed that the
restless mummy of the "Apewoman" was still in existence, stored in the base-
ment of the Institute of Forensic Medicine at the Rikshospitalet in Oslo. A
journalist sought permission to view and photograph the mummy, but the
chairman of the institute refused. It turned out that in 1979, the Oslo police
had been notified that some children had found a mummified human arm at
a dump in one of the suburbs of Oslo. The police found the rest of Julia
Pastrana's mummy in an abandoned caravan nearby. Strangely enough, they
did not notify the legal owner. Instead, they put the mummy into storage at

the Institute of Forensic Medicine. Julia Pastrana's Russian sedan chair is still kept at Lund's Tivoli, together with an archive of newspaper clippings, exhibition posters, and other material about her.

When I saw Julia Pastrana's mummy in 1990, it was standing on a small wooden board covered with fabric. The right arm had been torn off and lay in front of the mummy; the right side of the face had been torn open as well, and the eye on that side was missing. The Russian dancer's costume placed on the corpse in 1860 had been torn off by the thieves, and the mummy was completely unclothed apart from the remains of the original boots. The hairy growth was greatly diminished by the ravages of time, but the abnormal hairiness of the forehead was still evident and parts of the whiskers were also preserved. The skin was dark brown and parchmentlike. It seems that the lifelike appearance, which so astounded the nineteenth century spectators, has a simple explanation, for it was apparent that Sukolov had filled the arms and legs of the mummy with stuffing during the embalming process to prevent shrinkage. The preparation of the corpse was nevertheless very skillful, using minimal sutures.

Julia Pastrana's Diagnosis

For many years the majority of anthropologists and dermatologists have believed that Julia Pastrana suffered from congenital hypertrichosis lanuginosa (inherited increased hairiness with lanugo hair), a genetic disorder that is usually inherited as an autosomal dominant. Fewer than forty unrelated families with this form of hypertrichosis have been described. The bodies of the individuals with this disorder have been covered with long, soft, wavy hair, a persistence of the lanugo hair that covers the human fetus.

A remarkable early case was Petrus Gonzales, born in the Canary Islands in 1556, who had long, soft hair all over his body, like a Newfoundland dog. He was taken to the court of King Henry II of France as a curiosity, married there and had three children, who all inherited the same form of hairiness. At least one of them brought the disorder into a third generation.

A severe diminution in the number of teeth has often been observed in conjunction with congenital hypertrichosis lanuginosa. One example was Shwe-Maong, a Burmese man enveloped from head to foot in a mass of wavy hair. He and his descendants had only four or five teeth in either jaw, and where there were no teeth, the alveolar processes were missing also.

Several writers divide the individuals with congenital hypertrichosis lanuginosa into two groups. Nearly all cases, including Petrus Gonzales and

Shwe-Maong, are considered dog-faced. The entire face is covered with long hair and the entire body with thick, soft lanugo hair up to ten centimeters in length. The other subgroup, the monkey-faced, comprises only Julia Pastrana, her son, and Krao, a hairy girl from Bangkok who was widely exhibited in the 1890s. Their hair growth has more of a male pattern, with long whiskers and beard but shorter hair on the rest of the face. Furthermore, the body hair of Julia and Krao was short, hard, and jet-black.

The remaining hair on Julia Pastrana's mummy resembles normal terminal hair rather than lanugo. Microscopic examination of hair from the head and beard showed that all features were consistent with human terminal facial hair. Since replacement of lanugo with terminal hair never occurs in true congenital hypertrichosis lanuginosa, it is clear that Julia Pastrana did not have this disorder. But what was the cause of her strange deformities?

One important clue is the appearance of her teeth and jaws, which fascinated many of those who saw her during life. A curious rumor started by the exhibition pamphlets was that Pastrana had double gums and two sets of teeth; Charles Darwin, among others, had heard the story. On the other hand, some German anthropologists claimed that she had fewer than the normal number of teeth, seeming to support the diagnosis of hypertrichosis lanuginosa congenita. A pair of plaster casts of her upper and lower dentition, probably made in 1857, are kept in the Odontological Museum of the Royal College of Surgeons in London. They show greatly thickened alveolar processes, and it is difficult on the casts to distinguish cusps of teeth from nodules of overgrown gum. Even in life such nodules, covered with parakeratotic epithelium, could be whitish and thus easily confused with teeth, explaining the unfounded rumor of double rows of teeth.

The plaster casts and the accounts of Julia Pastrana during life agree that she suffered from severe gingival hyperplasia (overdevelopment of the gum). Whether this was congenital is not known, but it was present as early as 1855 when she was twenty-one years of age. It seems highly probable that the gingival hyperplasia was progressive and gradually buried any teeth that erupted. No observer would have been able to ascertain what was beneath the overgrown gum during her lifetime, but inasmuch as the jaw of the mummy is relatively undamaged, it is now possible to do so. A skull X-ray and panoramic X-rays of the mummy's jaws show that Julia Pastrana had a complete set of permanent (secondary) teeth, with the possible exception of the upper left lateral incisor.

With regard to Julia Pastrana's marked gingival hyperplasia and hypertrichosis with terminal hair, it is reasonable to propose that she suffered from another rare genetic disease, the autosomal dominant syndrome of congenital hypertrichosis and gingival hyperplasia, which is distinct from congenital hypertrichosis lanuginosa. It is also most likely that the individuals classified in the so-called monkey-faced group of congenital hypertrichosis lanuginosa in fact exhibit severe congenital hypertrichosis terminalis with gingival hyperplasia and facial deformity. Julia Pastrana represents an extreme case of congenital hypertrichosis (with terminal hair) and gingival hyperplasia, perhaps the most extreme of all time. Nothing even resembling her gross facial deformity — the thick lips, broad and flat nose, and large ears — has been seen in any later example of this syndrome. It is interesting that there is an autosomal dominant syndrome of gingival hyperplasia without hypertrichosis, which many consider a variant of the syndrome with hypertrichosis. It is tempting to speculate that there is variable expression, so that some individuals have only mild gingival hyperplasia, while others have gingival hyperplasia and less severe congenital hypertrichosis, and very few, such as Julia, a combination of severe gingival hyperplasia, extreme congenital hypertrichosis with terminal hair, and gross facial deformities.

The strange story of Julia Pastrana is not one to inspire confidence in human nature. Since the more trustworthy sources agree that in spite of her extraordinary appearance, Julia Pastrana was a normal, intelligent woman of gentle disposition, the true tragedy of her fate can scarcely be imagined. The callous exploitation of her during life is not unique in the annals of human deformities. Many other unfortunate "freaks," such as the Sicilian Fairy and the Elephant Man, suffered similar exhibition. But the macabre profiteering on her mummy for more than 110 years is quite without precedent. During her career in show business, before and after her death, Julia made such an impact on the world at large that her name remained a household word in large parts of Europe well into the twentieth century. In the standard anthropological text *The Living Races of Mankind*, published in 1900, there is a photograph of Julia Pastrana, allegedly taken during her life. Clearly though, it was in fact taken while her mummified remains were on tour in England in the 1860s. The very same photograph has also been used for more sinister purposes by certain American racist publications, which have identified it as the portrait of a hybrid between a black person and an ape.

As late as 1996, there was controversy in Norway about Julia Pastrana's mummy. Certain influential clergymen campaigned to have it cremated or destroyed; some even demanded that it be buried, with full ceremony, in the

Cathedral of Oslo. After they had presented a petition to the Norwegian home secretary, a government committee was instructed to inquire into the circumstances and determine the fate of Julia Pastrana's earthly remains. Several physicians and scientists believed it to be an act of bigoted vandalism to destroy the mummy, and they instead suggested that it be kept in a locked sarcophagus in a newly constructed museum of medical history in Oslo. Here, it would be available only for serious scientific study and would not be shown to the public or to vulgar sensation seekers. To the disappointment of the clergymen involved, this latter party gained the upper hand in the controversy, and Julia Pastrana's restless mummy is still with us, although, it is to be hoped, in the hands of more worthy guardians than those involved earlier.

While publishing the paperback, the following came to light:

According to information received in June 1998, Julia Pastrana's mummy is stored at an anatomical institute in Oslo. The mummy is not on show to the public, and probably never again will be.

Note on Sources

Spontaneous Human Combustion
Many of the older sources are mentioned in the text; two others, with extensive lists of references, are *Ausführlichen Darstellung und Untersuchung der Selbstverbrennungen* by J. H. Kopp (Frankfurt am Main, 1811) and *De la combustion humaine spontanée* by L. Delmas (Strasbourg 1867). The considerable secondary literature includes papers by C. D. Josephson (*Acta Medica Scandinavica* 65 [1926], 424–42), J. R. Oliver (*Bulletin of the History of Medicine* 4 [1936], 559–72), L. Adelson (*Journal of Criminal Law, Criminology and Police Science* 42 [1952], 793–809), M. Thomsen (*Burns* 5 [1978], 54–59, and *Dansk Medicinhistorisk Årbog* [1978], 126–58), and J. Bondeson (*Sydsvenska Medicinhistoriska Sällskapets Årsskrift* 24 [1987], 83–98). A recent valuable reexamination of the von Görlitz case is that of Dr. J. L. Heilbron (*Proceedings of the American Philosophical Society* 138 [1994], 284–316). Ethnological aspects are considered in papers by J. Scharffenberg (*Tidskrift for den Norske Lægeforening* 45 [1925], 470–75, 522–28), H.-J. Frey (*Schweizer Monatshefte* 44 [1964], 870–82), W. S. Walker (*Journal of Popular Culture* 16 [1982], 17–25), W. S. Walker and A. E. Uysal (*International Folklore Review* 4 [1986], 62–67), and S. Shaw (*Literature and Medicine* 14 [1995], 1–22). The modern debate on spontaneous combustion is well summarized in articles by D. J. Gee (*Medicine, Science, and the Law* 7 [1965], 37–38), J. Heymer (*New Scientist* 109(1508) [1986], 70–71, and *Fortean Times* 74 [1994], 29–32, 75, 37–40, and 79 [1995], 40–43), J. Randles and P. Hough (*Fortean Times* 63 [1992], 44–49), and J. Nickell and J. F. Fischer (*Skeptical Inquirer* 11 [1987], 352–57). The many papers on literary spontaneous combustions include those by R. D. McMaster (*Dalhousie Review* 38 [1958], 18–28), G. Perkins (*Dickensian* 60 [1965], 57–63), T. Blount (*Dickens Studies Annual* 1 [1970], 183–211), E. Gaskell (*Dickensian* 69 [1973], 25–35), A. Farrag (*Romance Notes* 19 [1978], 190–95), P. Denman (*Dickensian* 82 [1986], 131–41), L. B. Croft (*CLA Journal* 32 [1989], 335–47), and J. B. West (*Dickensian* 90 [1994], 125–29).

The Bosom Serpent
The considerable older literature is listed by J. H. Joerdens in his *Entomologie und Helminthologie des menschliches Körpers*, vol. 2 (Berlin, 1801), 129–44, and by A. A. Berthold in *Über den Aufenthalt lebender Amphibien im Menschen* (Göttingen, 1850). Important articles are those by E. G. Baldinger (*Neues Magasin für Aerzte* 1 [1779], 385–

401), Dr. Sander (*Wochenschrift für die gesammte Heilkunde* 39 [1834], 617–21), and M. W. Mandt (*Magazin für die gesammte Heilkunde* 53 [1839], 491–510). From the English-speaking world, a notable article appeared in *Kirby's Wonderful and Scientific Museum*, vol. 3 (London, 1802–20), 360–69, and another, by A. Stengel, in the *University of Pennsylvania Medical Bulletin* 16 (1903–4), 86–89. I myself published a considerable list of further references from the older medical literature (*Sydsvenska Medicinhistoriska Sällskapets Årsskrift* 27 [1990], 141–74). Nine important articles on the bosom serpent in literature and folklore are those by A. Jacoby (*Oberdeutsche Zeitschrift für Volkskunde* 6 [1932], 13–27), R. D. Arner (*Southern Folklore Quarterly* 35 [1971], 336–46), S. Bush Jr. (*American Literature* 43 [1971–72], 181–99), D. R. Barnes (*Journal of American Folklore* 85 [1972], 111–22), R. Jordan de Caro (*Journal of American Folklore* 86 [1973], 62–65), R. J. M. Rickard (*Fortean Times* 40 [1983], 17–18), G. Bennett (*Fabula* 26 [1985], 219–29), H. Schechter (*Georgia Review* 39 [1985], 93–108), and F. Cattermole-Tally (*Folklore* 106 [1995], 89–92).

The Riddle of the Lousy Disease

On phthiriasis among the ancients, see the papers by W. Nestle (*Archiv für Religionswissenschaft* 33 [1936], 246–69), H. Keil (*Bulletin of the History of Medicine* 25 [1951], 305–23), T. Africa (*Classical Antiquity* 1 [1982], 1–17), A. Keaveney and J. A. Madden (*Symbolae Osloenses* 57 [1982], 87–99), and J. Schamp (*L'Antiquité Classique* 50 [1991], 139–70). On medical aspects of the disease, see the original paper by A. C. Oudemans (*Zeitschrift für Parasitenkunde* 11 [1940], 145–98), which has an extensive list of references. His theories were commented on by R. Hoeppli in *Parasites and Parasitic Infection in Early Medicine and Science* (Singapore, 1959), 349–59, J. R. Busvine in *Insects, Hygiene, and History* (London, 1976), 102–6, 195–203, and J. Bondeson (*Sydsvenska Medicinhistoriska Sällskapets Årsskrift* 26 [1989], 197–236).

Giants in the Earth

A concise summary of the older literature can be found in *Mundus subterraneus* by Athanasius Kircher (Amsterdam, 1665), book 8, pp. 53–60, and the *Physica curiosa* by Gaspar Schottus (Herbipoli, 1662), 430–41. The original papers of Molyneux and Sloane were published in the *Philosophical Transactions of the Royal Society of London* (22 [1700], 487–508, and 35 [1728], 497–514). Several further documents of interest, including Sir Hans Sloane's own file on giants, are in the Archives of the Royal Society of London. An interesting account of the scandal associated with King Teutobochus is in the ninth volume of the *Variétés historiques* (Paris 1859), 241–59. Frank Buckland's *Curiosities of Natural History* (London, 1859), 2:32–43, and E. J. Wood's *Giants and Dwarfs* (London, 1868), 1–40, both contain sections on giants in the earth. Another important source is *Vorzeitliche Tierreste im deutschen Mythus, Brauchtum, und Volksglauben* by O. Abel (Jena, 1939), 97–114. More recent studies include the articles by Jean Céard (*Journal of Medieval and Renaissance Studies* 8 [1978], 37–

76), A. Schnapper (*Annales Economies, Sociétés, Civilisations* 41 [1986], 177–200), H. Fromm (*Deutsche Vierteljahrschrift für Literaturwissenschaft* 60 [1986], 42–59), and D. Levin (*William and Mary Quarterly* 45 [1988], 751–70), as well as the important book *There Were Giants in Those Days* by W. E. Stephens (Lincoln, 1989). The Cardiff Giant is described in an anonymous contemporary pamphlet titled *The Cardiff Giant Humbug* (Fort Dodge, 1870) and in the articles by B. Franco (*New York History* 50 [1969], 421–40), and R. Delorme (*Historia*, no. 428 [1982], 122–27), and F. D. Schwartz (*American Heritage* 45, [1994], 31–32). The Cardiff Giant also has two homepages on the Internet: www.cardiffgiant.com and www.lhup.edu/~dsimanek/cardiff.htm.

Apparent Death and Premature Burial

The major eighteenth-century works on this subject were the various editions of Winslow and Bruhier, as well as *Lettres sur la certitude des signes de la mort* by A. Louis (Paris, 1752). C. W. Hufeland wrote many works on apparent death, of which *Der Scheintod* (Berlin, 1808) was one. On the German waiting mortuaries, see *Das Leichenhaus in Weimar* by C. Schwabe (Leipzig, 1834). The many pamphlets by H. Le Guern were typical of the nineteenth-century agitation to prevent premature burial, but the book *Traité des signes de la mort* by E. Bouchut (Paris, 1849, and several later editions) attempted to call the French nation to order. On the late nineteenth-century debate, see the works of Hartmann, Walsh, and Tebb and Vollum quoted in the text. Two popular nineteenth-century American reviews were those by G. E. Mackay (*Popular Science Monthly* 16 [1880], 389–97) and M. Dana (*Arena* 17 [1897], 935–39). The American introduction of Count de Karnice-Karnicki's coffin was described by E. Camis (*Medico-Legal Journal* 17 [1900], 296–99). Important secondary literature includes the thesis "Die Angst vor dem Scheintod in der 2. Hälfte des 18. Jahrhunderts," by Dr. Martin Patak (*Zürcher Medizingeschichtliche Abhandlungen*, no. 44 [Zürich, 1967]), the book *The Hour of Our Death* by Philippe Ariès (London, 1983), and the articles by H. Dittrick (*Journal of the History of Medicine* 3 [1948], 161–71), L. G. Stevenson (*Bulletin of the History of Medicine* 49 [1975], 482–511), T. K. Marshall (*Medicolegal Journal* 35 [1966], 14–24), N. Roth (*Medical Instrumentation* 14 [1980], 322), and M. Alexander (*Hastings Center Report* 10 [1980], 25–31). Two interesting Dutch articles are those by E. M. Terwen-Dionisius (*Jaarboek die Haghe, 1986*, 66–99) and C. van Raak (*Maatstaf* 42[2] [1994], 32–38). For an eccentric modern view of the subject, see *Buried Alive* by Dr. Péron-Autret (London, 1983). For summaries of modern newspaper stories of premature burial, see the articles by P. Sieveking (*Fortean Times* 49 [1987], 55–60, and 63 [1992], 36–38); this subject has also been quite extensively discussed in the alt.folklore.urban newsgroup on the Internet. The literary sources on premature burial have been examined by W. T. Bandy (*American Literature* 19 [1947], 167–68) and G. Kennedy (*Studies in the American Renaissance* [1977], 165–78).

Mary Toft, the Rabbit Breeder

The original sources are mentioned in the text. Many of them are gathered in the Samuel Merriman volume in the Library of the Royal Society of Medicine (MSS. 265) and in a volume titled *Tracts Related to Mary Toft* in the British Library. Four important later articles are those by A. J. Wallace (*Liverpool Medicochirurgical Journal* 62 [1912], 254–71), J. Avalon (*Aesculape* 22 [1932], 274–81), S. A. Seligman (*Medical History* 5 [1960], 349–60), and G. Leslie (*Eighteenth Century* 27 [1986], 269–86). James Douglas's part in the scandal is further elucidated in *James Douglas of the Pouch* by K. B. Thomas (London, 1964), 60–69, 139–40. The caricatures celebrating the scandal are listed in the *Catalogue of Prints and Drawings in the British Museum*, div. 1, vol. 2, pp. 633–50). Hogarth's drawings are studied in detail in R. Paulson's *Hogarth: His Life and Times*, vol. 1 (London, 1971), 168–70, and in the excellent papers by D. Todd (*Eighteenth-Century Studies* 16 [1982], 26–46, and *Studies in Bibliography* 41 [1988], 247–67). Of the local history sources consulted, *Godalming Parish Church* by A. Botts (Godalming, 1987) deserves particular mention.

Maternal Impressions

The voluminous original literature on this subject, enumerating several German and French doctoral dissertations as well as hundreds of scientific papers and case reports, is listed in the first three series of the Index Catalogue of the Surgeon General's office. Three modern theses are those of F. Kahn, "Das Versehen der Schwangeren in Volksglaube und Dichtung" (University of Berlin, 1913), O. Bossert, "Ein Beitrag zur Lehre vom Versehen" (University of Heidelberg, 1914), and B. Stokvis, "Het Verzien in de Zwangerschap" (University of Leiden, 1940). A particularly well-written account is that of J. W. Ballantyne in his *Antenatal Pathology and Hygiene*, vol. 2 (London, 1904), 105–28. Some important historical papers are those by J. D. Herholdt (*Det Kongelige Danske Videnskabernes Selskabs Nat. & Math. Afd.* 4 [1829], 257–320), G. J. Fisher (*American Journal of Insanity* 26 [1870], 241–95), J. C. Huber (*Friedrichs Blätter für gerichtliches Medizin* [1886], 321–24), and J. W. Ballantyne (*Edinburgh Medical Journal* 36 [1890–91], 624–35, and 37 [1891–92], 1025–34). The role of maternal impression in folklore has been discussed by A. F. Fife in *American Folk Medicine*, ed. W. D. Hand (Berkeley, 1976), 273–83, and in the articles by A. McLaren (*New Scientist* 22 [1964], 97–100), J. H. Fearn and H. Parlin (*Medical Journal of Australia* 2 [1971], 1123–26), W. C. Shaw (*British Journal of Plastic Surgery* 34 [1981], 237–46), and H. W. Roodenburg (*Journal of Social History* 21 [1988], 701–16). Literary aspects are discussed by G. S. Rousseau in *Bicentennial Essays Presented to Lewis M. Knapp*, ed. G. S. Rousseau and P. G. Boucé (New York, 1971), 79–109, J. C. Garver (*Saeculum* 33 [1983], 287–311), M. Carlson (*Proceedings of the Modern Language Association* 100 [1985], 774–82), and P. G. Boucé in *Sexual Underworlds of the Enlightenment*, ed. G. S. Rousseau and R. Porter (Chapel Hill, N.C., 1988), 86–100.

Tailed People

The two papers by Dr. Max Bartels (*Archiv für Anthropologie* 13 [1881], 1–41, and 15 [1884], 45–131) review the idea of tailed human races most thoroughly, and the articles by O. Schaeffer (*Archiv für Anthropologie* 20 [1891], 189–224) and C. Hennig and A. Rauber (*Virchows Archiv* 105 [1886], 83–109) give supplementary details of early cases. Historical reviews about tailed people have been written by S. Baring-Gould, *Curious Myths of the Middle Ages* (London, 1868), 145–60, and by O. Sherwin (*Journal of the History of Medicine* 13 [1958], 435–68), J. C. K. Rijsbosch (*Archivum Chirurgicum Neerlandicum* 29 [1977], 261–68), and F. Ledley (*New England Journal of Medicine* 306 [1982], 1212–15). Some important clinical papers about tailed children have been written by R. G. Harrison (*Johns Hopkins Hospital Bulletin* 12 [1901], 96–101), A. H. Dao and M. G. Netsky (*Human Pathology* 15 [1984], 449–53), and S. Chakrabortty et al. (*Journal of Neurosurgery* 78 [1993], 966–69). Linnaeus's "Anthropomorpha" has been reprinted in Swedish (*Valda Avhandlingar av Carl von Linné*, no. 21), and the documents concerning the woman with a horse's tail are in the Archives of the Royal Swedish Academy of Sciences (Sekr. arkiv 8:2).

Three Remarkable Specimens in the Hunterian Museum

Two good modern biographies of John Hunter are *John Hunter* by J. Dobson (Edinburgh, 1969) and *John Hunter (1728–1793)* by G. Qvist (London, 1981); furthermore, *The Case Books of John Hunter, FRS* were published in 1993. On the Irish Giant, the main sources are *Giants and Dwarfs* by E. G. Wood (London, 1868), 157–65; the *Descriptive Catalogue of the Physiological Series in the Hunterian Museum*, vol. 2 (London, 1971), 199–206; and the biography of Patrick Cotter, *The Irish Giant* by G. Frankcom and H. Musgrave (London, 1976). Articles on the Giant include those by R. M. Bergland (*Journal of Neurosurgery* 23 [1965], 265–69), N. H. McAlister (*Canadian Medical Association Journal* 111 [1974], 256–57), and A. M. Landolt and M. Zachmann (*Lancet*, no. 1 [1980], 1311–12). On the Two-Headed Boy of Bengal, see the article by J. Bondeson and E. Allen (*Surgical Neurology* 31 [1989], 426–34) and its references. On the Sicilian Fairy, see the papers by E. Home (*Philosophical Transactions* 115[1] [1825], 66–80), H. Gilford (*Medico-Chirurgical Transactions* 85 [1902], 305–59), and J. Dobson (*Annals of the Royal College of Surgeons* 16 [1955], 268–72), as well as the book *Bird-Headed Dwarfs* by H. P. G. Seckel (Basel, 1960). The paper by J. Bondeson (*American Journal of Medical Genetics* 44 [1992], 210–19) has an extensive list of references. I thank Elizabeth Allen, curator of the Hunterian Museum, and Ian F. Lyle, librarian of the Royal College of Surgeons, for valuable help.

The Strange Story of Julia Pastrana

Two adequate reviews on congenital hypertrichosis are those by W.-R. Felgenhauer (*Journal de Génétique Humaine* 17 [1969], 1–44) and T. W. Nowakowski and A. Scholz (*Der Hautarzt* 28 [1977], 593–99). Julia Pastrana has also had a doctoral

250   dissertation written about her, titled "Über Trichosen, besonders die der Julia Pastrana I" (University of Bonn, 1917), by Dr. J. Fuchs. The articles by A. E. W. Miles
(*Proceedings of the Royal Society of Medicine* 67 [1974], 160–64), and J. Bondeson and
A. E. W. Miles (*American Journal of Medical Genetics* 47 [1993], 198–212) contain a
complete bibliography of the relevant literature.

Jan Bondeson is a physician specializing in rheumatology and internal medicine. He holds a Ph.D. in experimental medicine and works at a major research institute in London.